Astrophysics and Space Science Proceedings
Volume 31

For further volumes:
http://www.springer.com/series/7395

Stellar Pulsations

Impact of New Instrumentation and New Insights

Editors

J. C. Suárez
R. Garrido
L. A. Balona
J. Christensen-Dalsgaard

Springer

Editors
J. C. Suárez
R. Garrido
CSIC
Instituto de Astrofísica de Andalucía
Granada
Spain

L. A. Balona
South African Astronomical Observatory
Cape Town
South Africa

J. Christensen-Dalsgaard
Department of Physics and Astronomy
Aarhus University
Aarhus
Denmark

ISSN 1570-6591 ISSN 1570-6605 (electronic)
ISBN 978-3-642-29629-1 ISBN 978-3-642-29630-7 (eBook)
DOI 10.1007/978-3-642-29630-7
Springer Heidelberg New York Dordrecht London

Library of Congress Control Number: 2012943346

Printed on acid-free paper

Springer is part of Springer Science+Business Media (www.springer.com)

This book is dedicated to Arthur N. Cox.

Foreword

It has been a great honor and responsibility for us to organize the 20th edition of the prestigious Stellar Pulsation Conference Series, held for the first time in Spain. As well, it has been a privilege to count among us with some of the most reputed researchers who have written the foundation of this scientific discipline.

We wish all the best to the organizers of the next conference, 21st of the series, which will be held in Wroclaw, Poland (2013).

Granada, Spain
September 2011

Juan Carlos Suárez
Rafael Garrido

Preface

This conference is the 20th edition of the Stellar Pulsation Series (known as the "Los Alamos" Series) held for the first time in Spain:

1971	Los Alamos, USA
1974	Goddard Space Flight Center, USA
1976	Los Alamos, USA
1978	Goddard Space Flight Center, USA
1980	Los Alamos, USA
1982	Boulder, USA
1984	Toronto, Canada
1986	Los Alamos, USA
1988	University of Nebraska, USA
1990	Bologna, Italy
1992	Victoria, Canada
1995	Cape town, South Africa
1997	Los Alamos, USA[1]
1999	Budapest, Hungary
2001	Leuven, Belgium
2003	Christchurch, New Zealand
2005	Rome (Monte Porzio Catone), Italy[2]
2007	Vancouver, Canada
2009	Los Alamos (Santa Fe), USA
2011	Granada, Spain[3]

The number of participants (125) and the quality of the contributions to be found in these proceedings are a measure of the interest of the scientific community in the topics for discussion proposed by the Scientific Organizing Committee. The title

[1] http://www-xdiv.lanl.gov/XTA/pulsation/

[2] http://www.mporzio.astro.it/~step/

[3] http://stellarpulsations.iaa.es/

of the first session, “Poor agreement between observed predicted frequencies,” illustrates the intention of the organizers to stimulate discussion of controversial views by specifically selecting invited contributions dealing with these issues.

These proceedings include topics which have not been presented in previous issues of this series. Among these are contributions dealing with the impact of stellar evolution on pulsations and vice versa and on pulsations of host stars with planets. In addition, a special session dedicated to new space- and ground-based instrumentation has been included.

The success of a scientific meeting can be evaluated by the number of problems that has been resolved and, inevitably, the number of new questions which arises in the search for answers. A very pertinent question is, of course, whether or not asteroseismology is delivering on its promises. An answer to this question can be found in the excellent review of the Conference by Joyce Guzik (see “Closing Summary,” in these proceedings).

September 2011

Juan Carlos Suárez
Rafael Garrido
Luis A. Balona
Jørgen Christensen-Dalsgaard

Acknowledgements

First of all, we acknowledge the entire local and scientific organizing committees for their support and dedication to the conference. We acknowledge the financial support from the Spanish Ministry of Science and Innovation through the project AYA2010-20982-C02-01.The Instituto de Astrofísica de Andalucía (CSIC) is also acknowledged for its support and advisory tasks. We acknowledge El Corte Inglés (Granada) travel agency for their competent work as organization managers. As well, the hospitality and attention by the managers of the Abades Nevada Palace is acknowledged. Finally, we acknowledge Dr. Joyce Guzik for her generosity and dedication in preparing the closing talk with the summary of the conference and Silbia López de la Calle for her proficient work in the design of logos and posters.

Contents

Part I
The Poor Agreement Between Observed and Predicted Frequencies

Stellar Pulsations and Stellar Evolution: Conflict, Cohabitation, or Symbiosis?

Achim Weiss

Abstract While the analysis of stellar pulsations allows the determination of current properties of a star, stellar evolution models connect it with its previous history. In many cases results from both methods do not agree. In this review some classical and current cases of disagreement are presented. In some cases these conflicts led to an improvement of the theory of stellar evolution, while in others they still remain unsolved. Some well-known problems of stellar physics are pointed out as well, for which it is hoped that seismology—or in general the analysis of stellar pulsations—will help to resolve them. The limits of this symbiosis will be discussed as well.

1 Living Together

The properties of stellar pulsations—appearance, periods, multiperiodicity—depend on the current state of a star. They are sensitive to the equilibrium structure and its reaction with respect to perturbations. Using pulsations, radial and non-radial, one may determine global properties of the star, such as mean density, gravitational acceleration, and mass-to-light ratio, but also gain access to the inner structure (more specifically, to the driving regions or mode cavities). Period changes, detectable through phase changes observed over longer times, allow to measure the direction and speed of evolution.

From the point of view of stellar evolution theory, a pulsating star is the current end product of a long history of evolution, containing both global and initial conditions (such as mass and composition), and all the physics of the evolution up to the present time. Stellar evolution theory determines the full history and

A. Weiss (✉)
Max-Planck-Institut für Astrophysik, Karl-Schwarzschild-Str. 1, 85748 Garching, Germany
e-mail: aweiss@mpa-garching.mpg.de

Weiss, A.: *Stellar Pulsations and Stellar Evolution: Conflict, Cohabitation, or Symbiosis?*. Astrophys Space Sci Proc. **31**, 3–10 (2013)
DOI 10.1007/978-3-642-29630-7_1,

complete structure of the observed object at any time prior to the observation. Given the complexity of the constitutional physics, a mismatch between the pulsationally determined structure or position in the Hertzsprung–Russell-diagram (HRD), is not totally surprising, but may depend on physics not directly related to the present state of the star. This is what makes the comparison between stellar pulsations and evolutionary models sometimes so complicated.

1.1 Conflicts

A classical, and still existing conflict is that of the mass of Cepheids. Since the 1970s it was found in the work by Christy [12], Stobie [45], Rodgers [38], Fricke et al. [18], and others that Cepheid pulsation periods require, at given luminosity, a stellar mass lower by at least 10%, and up to a factor of two, than that which is obtained from stellar evolutionary tracks matching the observed HRD position. The conflict was partially resolved with the new generation of Rosseland opacities in the 1990s [39, 41], which influence the pulsation periods, but still exists at an average level of 20%.

There are two possible resolutions, both related to poorly understood aspects of stellar evolution. Either the present mass of the Cepheid is lower than its initial mass, or convective overshooting on the main sequence has taken place, which leads to higher luminosity, also during the Cepheid (helium burning) phase. Mass loss rates of $\approx 10^{-11}\ M_\odot$/year, which would be the level predicted by the standard Reimers formula [37], are insufficient for the former explanation, but a pulsation-driven mass loss proposed by [30] may reach (due to shocks) levels of $10^{-7}\ M_\odot$/year. It may be able to remove 5–10% of the mass and thus resolve this *Cepheid mass discrepancy* at least partially [29]. Nevertheless, these models additionally needed core overshooting to reproduce Cepheid masses from their evolutionary tracks.

A particular interesting case is that of the detached eclipsing binary system OGLE-LMC-CEP0227 [35], in which component A is a Cepheid of $4.14 \pm 0.05\ M_\odot$, and component B is a star of almost identical mass, but cooler than the red edge of the instability strip. Stellar models by [11], computed with the same amount of overshooting from convective cores that is needed to reproduce the CMD morphology at the turn-off of open clusters, are able to reproduce the positions of both components in the $T_{\rm eff}$–R diagram for identical ages of $\approx$ 154 Myr. Mass loss according to the Reimers formula was included in the models, but did reduce masses (initially 4.14 and 4.21 $M_\odot$ for A and B) by only 0.03 $M_\odot$ in both cases.

In Fig. 1 I show my own solution for the same object. The stellar models were computed with our stellar evolution code GARSTEC, described in [49]. The physical ingredients are very similar to those of [40]. In particular, convective overshooting is included as well. The chemical composition was identical to that in [40], appropriate for the LMC. Again, the masses needed to reproduce R at the same age (156 Myr) are within the errors of the dynamical masses (4.14 $M_\odot$, resp. 4.24 $M_\odot$). As in [11] component B is slightly too bright compared to the tracks. A significant difference, however, exists in the loops, which here are much more

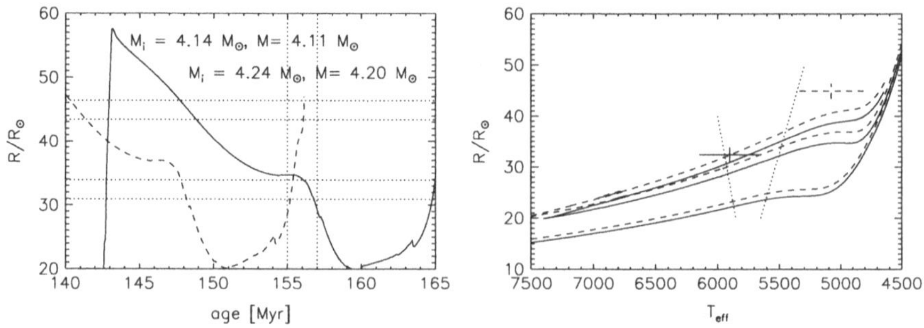

Fig. 1 Models for OGLE-LMC-CEP0227, computed with my own code GARSTEC, including convective overshooting from all convective boundaries and Reimers mass loss. *Left*: radius R of components A (Cepheid, *solid line*) and B vs. time. Observed radii are indicated by *dashed horizontal lines*; the age range obtained from this calculation by *vertical* ones. Initial and present masses are given, too. *Right*: Tracks for the same models in the T_{eff}–R diagram, along with the Cepheid instability strip (*dotted lines*) and the observed positions. Line-styles of object and track correspond

extended than in the models of [11]. This is probably due to the fact that in my models overshooting is also applied from the bottom of the convective envelope [10]. In any case, this example very convincingly indicates that the Cepheid mass discrepancy can and has to be resolved by properly taking into account overshooting from convective cores.

A more recent conflict concerns the Standard Solar Model, in relation with the latest *solar abundances* [2]. The seismic structure of the Sun, obtained from the analysis of its non-radial oscillations, is extremely well-determined. For example, the depth of the convective envelope is $0.713 \pm 0.001\ R/R_\odot$ and the present envelope helium content 0.249 ± 0.003. The sound speed profile is known with an r.m.s. error of only 0.003 [4]. With the new abundance determinations, which led to a reduction in C, N, and O of up to 30% (and thus to $Z/X = 0.017$ compared to the previous value of 0.023), the agreement between the solar model and these values is much worse [3]. In spite of many attempts, the seismically inferred values and the solar model prediction could not be brought back to the previous agreement, which was, for example for the convective envelope depth, nearly perfect. The general agreement is that—assuming the abundances are indeed correct—something is wrong with the models or the physics used therein.

1.2 Cohabitation

With *cohabitation* I mean that both pulsation and stellar evolution theory share problems with the same type of stars, without a clear route to a common resolution being visible. A first example are RR Lyr stars, which are Horizontal Branch (HB) objects. Evolution on the HB has its own uncertainties. One particular problem with

RR Lyr stars is the Oosterhoff dichotomy [33], which is the apparent presence of two different populations of RR Lyr stars, characterized by different mean periods, separated by a gap in the period distribution. The relative fraction of RR Lyr variables within any Globular Cluster (GC) varies, giving rise to denoting GCs as being of type Oosterhoff I or II, which appears to depend on the cluster metallicity. One explanation dating back to [47] is based on a mode hysteresis that could result from changing directions of evolution through the instability strip. Incidentally, the evolution from the zero-age HB, and the occurrence of a blue-red-type evolution depend on the treatment of semiconvection in core helium-burning low-mass stars [15], and is therefore rather uncertain. Alternatively, different helium contents have been proposed [40] for being the reason for the Oosterhoff dichotomy. At the present time, photometric and spectroscopic observations of GCs find increasing evidence for the presence of sub-populations [8], which may or should differ also in the helium content (see, e.g. [45] for the most prominent case of ω Cen). At least for the first suggestion period changes may help to elucidate the situation. Otherwise, the mentioned problems in pulsation and evolution theory may or may not be connected.

The second example is *rotation*, which, in stars, leads to a change of the effective gravitational force and to non-spherical objects. Its major consequence for stellar evolution is often the occurrence of flows and mixing, and a change in the chemical stratification. It may interact with mass loss. So far, there are complete one-dimensional stellar models taking rotation into account in an approximate way (see [25] for a review), or multi-dimensional hydro-simulations of specific situations (see these proceedings and [19] for the specific case of RR Lyr stars). None of these approaches is able to solve the problems caused by rotation completely.

The problems caused for pulsation theory are also manifold: rotation leads to mode splitting, mode coupling, and a more difficult mode identification (although he mode splitting might also be helpful in identifying modes, provided the splitting is understood theoretically). Several of these issues, and how the effect on pulsation periods is treated will be the subject of other contributions, e.g. by Ouazzani, Shibahashi, in this meeting. Examples for successful detections of rapid core rotation are the cases of ν Eri and 12 Lac [16] and HD129929 [1].

1.3 Symbiosis

The perfect example for a successful symbiosis between pulsation and stellar structure theory is given by the mismatch of bump and beat Cepheid period ratios, and the apparent stability of β Cepheids. In his paper of 1982 [42], Simon summarized the problems and the need for an increase of metal opacities by a factor of 2–3 for $T > 10^5$ K. This triggered the "new" opacity calculations by the Livermore group and the Opacity Project. The results matched very well the predicted increase, and led to the resolution of the problem. The increase in opacity is now well-known as the Z- or *metal-bump*.

The new opacities also influenced the stellar models, and not only their pulsational properties. For example, the mass discrepancy in beat Cepheids was reduced, both from changes in period ratios [28] and by the fact that the models get less luminous [6]. The sound speed of the solar model differed less from seismic values, too. On the other hand, the extension of the Cepheid loops was reduced and models below 6 $M_\odot$ in some calculations did not reach the instability strip at all (the solution might be the appearance of "undershooting" from the convective envelope mentioned in Sect. 1.1). Generally, however, stellar models agreed much better with observations, in particular in their effective temperatures.

A second case of symbiotic developments in both fields is convection, treated beyond the classical, non-local, time-independent mixing-length theory (MLT). The interaction between pulsations and convection has for decades triggered the development of time-dependent variants of the MLT (see, e.g., [20, 44, 48]), which have been quite successful in reproducing the red edge of the instability strip, the Hertzsprung progression and the period–colour–luminosity relation for Cepheids, and more (see [7] for more details). On the side of stellar modelling, 2- and 3d-hydrodynamical calculations have been performed by Stein and Nordlund [31], Asplund [32], Collet [13, 14], Steffen, Freytag, and Ludwig [24], Porter and Woodward [36], and more. Muthsam (this volume) will present sample results about the potential of such calculations. Such models yield $T - \tau$-relations in the convective envelope and atmosphere, allow the "calibration" of the MLT parameter, and give insight into mode excitations. Geraux and Dupree [19, and this volume] have started to compute convective 3d-models with pulsations.

On the more empirical side, stellar evolution modelers try to determine the amount of overshooting from fitting their models and isochrones to observed objects, such as binary systems and open cluster CMDs. Examples can be found in [26]. Overshooting not only helps to remove the Cepheid mass discrepancy, but also affects period spacings in γ Dor stars [27].

2 A Stellar Modeler's Wish List

As in stellar evolution theory many physical aspects are treated in an approximate, parameterized, and altogether unsatisfying way, we need additional information from other fields to restrict our "freedom" in modelling stars. This section summarizes[1] quantities required for improved, more accurate and reliable models.

- *Mass*: one of the fundamental quantities for stellar evolution, needed for modelling field stars, but also checking isochrones (mass at turn-off). See the contributions by de Ridder, Roxburgh, Grigahčene (and by others) in this volume and, for example, [21], for ways how to obtain it for pulsating stars.

[1] Based on the second part of the talk in a much abbreviated form, due to the lack of space.

- *Mixed cores*: refers mainly to size of convective cores (see Cepheids), which determine luminosity and later phases of evolution. Asteroseismology offers promising ways to determine this [43].
- *Composition profiles:* the generalization of the mixed cores; in particular steep gradients in composition; relevant, e.g., for the position of the bump in GC CMDs. A successful example is the determination of pulsating white dwarf envelope mass, core composition, etc. (see [22] for some examples)
- *Solar abundances:* tests independent of the Sun are needed; examples are β Cep, ζ Oph and SPB stars, where the new Asplund abundances give better agreement with observations (e.g. [50]).
- *Helium abundance:* has become a decisive ingredient for stellar evolution in connection with GC sub-populations, and the main-sequence broadening [9].
- *Rotation:* determination of rotation profile in stars will give restrictions to simple models and show, where its effects on stellar evolution are important.
- *Diffusion:* is known to be essential for the Solar Model, but unclear, how effective in other stars. Analysis of pulsations of white dwarfs, for example, restrict chemical composition in such a way that diffusion is needed to explain this [17].
- *Speed of evolution:* period changes may relate to evolutionary changes. Important, e.g., for Cepheid loops; here, observed period changes do not match second and third crossing speeds [46]. Period changes have also been detected in white dwarfs (G117-B15A, [23]), attributed to the cooling of the star, and hence possibly constraining masses of exotic particles, such as axions [5], which influence the cooling process. Another very impressive example is V725 Sgr [34], which changed its period from 12 d in 1926 to now more than 90 d. It is believed to be an AGB star caught during a thermal flash that moves the star through the instability strip.

3 Conclusions

Pulsating stars are found all over the HRD, thereby covering stars with different physical effects being decisive for their structure and evolution. Examples are rotation and convective core sizes on the main sequence, semiconvection on the HB, superadiabatic convection in red giants, or radiative levitation in hot dwarfs and subdwarfs. Modelling pulsating stars can thus deliver valuable hints how to deal with physics that is not well understood. In the present era of massive surveys and space missions, the two communities should work together more closely than ever.

Acknowledgements The author wishes to thank the organizers for the invitation, the financial support, and for a truly stimulating meeting. The editors have helped to improve this paper by their very careful reading and constructive commenting.

References

1. Aerts, C., Thoul, A., Daszyńska, J., et al.: Science **300**, 1926 (2003)
2. Asplund, M., Grevesse, N., Sauval, A.J., Scott, P.: Ann. Rev. Astron. Astrophys. **47**, 481 (2009)
3. Bahcall, J.N., Basu, S., Pinsonneault, M., Serenelli, A.M.: Astrophys. J. **618**, 1049 (2005)
4. Basu, S., Pinsonneault, M.H., Bahcall, J.N.: Astrophys. J. **529**, 1084 (2000)
5. Bischoff-Kim, A., Montgomery, M.H., Winget, D.E.: Astrophys. J. **675**, 1512 (2008)
6. Bohm-Vitense, E., Evans, N.R., Carpenter, K., Morgan, S., Beck-Winchatz, B., Robinson, R.: Astron. J. **114**, 1176 (1997)
7. Buchler, J.R.: In: Szabados, L., Kurtz, D.W. (eds.) The Impact of Large-Scale Surveys on Pulsating Star Research, ASP Conv. Ser., vol. 203, pp. 343–355 (2000)
8. Carretta, E., Bragaglia, A., Gratton, R., D'Orazi, V., Lucatello, S.: Astron. Astrophys. **508**, 695 (2009)
9. Casagrande, L., Flynn, C., Portinari, L., Girardi, L., Jimenez, P.: Mon. Not. Roy. Ast. Soc. **382**, 1516 (2007)
10. Cassisi, S.: In: Kurtz, D.W., Pollard, K.R. (eds.) Variable Stars in the Local Group, ASP Conv. Ser., vol. 310, pp. 489–497 (2004)
11. Cassisi, S., Salaris, M.: Astrophys. J. Lett. **728**, 43 (2011)
12. Christy, R.F.: Quart. J.R.A.S **9**, 13 (1968)
13. Collet, R.: Phys. Scr. **133**, 014004 (2008)
14. Collet, R., Magic, Z., Asplund, M.: eprint arXiv:1110.5475 (2011)
15. Dorman, B., Rood, R.T.: Astrophys. J. **409**, 387 (1993)
16. Dziembowski, W.A., Pamyatnykh, A.A.: Mon. Not. Roy. Ast. Soc. **385**, 2061 (2008)
17. Fontaine, G., Brassard, P., Charpinet, S., Chayer, P.: Mem. S.A.It **77**, 49 (2006)
18. Fricke, K., Stobie, R.S., Strittmatter, P.A.: Astrophys. J. **71**, 593 (1972)
19. Geroux, Ch.M., Deupree, R.G.: Astrophys. J. **731**, 18 (2011)
20. Gough, D.O.: Astrophys. J. **214**, 196 (1977)
21. Hekker, S., Basu, S., Stello, D., et al.: Astron. Astrophys. **530**, A100 (2011)
22. Kawaler, S.D.: Astron. Soc. Pacific Conf. Ser. **83**, 81 (1995)
23. Kepler, S.O., Costa, J.E.S., Castanheira, B.G., et al.: Astrophys. J. **634**, 1311 (2005)
24. Ludwig, H.-G., Freytag, B., Steffen, M.: Astron. Astrophys. **346**, 111 (1999)
25. Maeder, A., Meynet, G.: Rev. Mod. Phys. **84**, 25 (2012)
26. Magic, Z., Serenelli, A., Weiss, A., Chaboyer, B.: Astrophys. J. **718**, 1378 (2010)
27. Miglio, A., Montalbán, J., Noels, A., Eggenberger, P.: Mon. Not. Roy. Astr. Soc. **386**, 1487 (2008)
28. Moskalik, P., Buchler, J.R., Marom, A.: Astrophys. J. **385**, 685 (1992)
29. Neilson, H.R., Cantiello, M., Langer, N.: Astron. Astrophys. **529**, L9 (2011)
30. Neilson, H.R., Lester, J.B.: Astrophys. J. **684**, 569 (2008)
31. Nordlund, Å, Stein, R.F.: Comp. Phys. Comm. **59**, 119 (1990)
32. Nordlund, Å, Stein, R.F., Asplund, M.: Living Rev. Sol. Phys. **6**, 1 (2008)
33. Oosterhoff, P.T.: Obs. **62**, 104 (1939)
34. Percy, J.R., Molak, A., Lund, H., et al.: Pub. Astron. Soc. Pacific **118**, 805 (2006)
35. Pietrzyński, G., Thompson, I.B., Gieren, W., Graczyk, D., Bono, G., Udalski, A., Soszyski, I., Minniti, D., Pilecki, B.: Nature **468**, 542 (2010)
36. Porter, D.H., Woodward, P.R.: Astrophys. J. Supp. Ser. **127**, 159 (2000)
37. Reimers, D.: Mem. Soc. Roy. Sci. Liège **8**, 369 (1975)
38. Rodgers, A.W.: Mon. Not. Roy. Astr. Soc. **151**, 133 (1970)
39. Rogers, F.J., Iglesias, C.A.: Astrophys. J. Suppl. Ser. **79**, 507 (1992)
40. Sandage, A.: Astrophys. J. **248**, 161 (1981)
41. Seaton, M.J, Yan, Y., Mihalas, D., Pradhan, A.K.: Mon. Not. Roy. Astron. Soc. **266**, 805 (1994)
42. Simon, N.R.: Astrophys. J. Lett. **260**, 87 (1982)
43. Silva Aguirre, V., Ballot, J., Serenelli, A.M., Weiss, A.: Astron. Astrophys. **529**, 63 (2011)
44. Stellingwerf, R.F.: Astrophys. J. **262**, 330 (1982)

45. Stobie, R.S.: Mon. Not. Roy. Ast. Soc. **144**, 485 (1969)
46. Turner, D.G., Abdel-Sabour Abdel-Latif, M., Berdnikov, L.N.: Mem. S.A.It. **77**, 152 (2006)
47. van Albada, T.S., Baker, N.: Astrophys. J. **185**, 477 (1973)
48. Unno, W.: Pub. Astron. Soc. Jpn. **19**, 140 (1967)
49. Weiss, A., Schlattl, H.: Astrophys. Space Sci. **316**, 99 (2008)
50. Zdravkov, Th., Pamyatnykh, A.A.: J. Phys. Conf. Ser. **118**, 012079 (2008)

2D Simulations of the Interaction Between Radial Pulsation and Convection

Chris Geroux and Robert G. Deupree

Abstract We have developed a code which can perform simulations of the interaction of convection and large-amplitude stellar pulsation in 2D and 3D, with an algorithm which uses the interior mass as the radial independent variable in all cases. Results and comparisons of our multidimensional simulations will be presented.

1 Introduction

1D time-dependent convective models have been used for a long time to model pulsating stars with time-dependent mixing-length theory convection. These models still have difficulties reproducing observed features of pulsating RR Lyraes, especially near the red edge of the instability strip. Gehmeyr [5] who used the Stellingwerf [11] formalism for time dependent convection, added an adaptive grid to better resolve the steep temperature gradients in the ionization region. He had difficulties reproducing the observed velocity amplitude–light amplitude relations and the light amplitude–rise time relations. Bono [1] mentioned that while including time-dependent convection does lower the pulsation amplitudes relative to the purely radiative models they are still not as low as is observed near the red edge. More recently Marconi [8] modelled RR Lyrae stars in M3 and found that two different mixing-length parameters were required to match both the red and blue edges of the instability strip. Later Marconi [9] reiterated that observed light curves are still not well reproduced by models near the red edge of the instability strip.

Deupree [3] approached the problem of convection in a fundamentally different way by following the large-scale convective flow patterns directly in 2D. The convective motions in his calculations arise naturally in unstable regions from very

C. Geroux (✉) · R.G. Deupree
Saint Mary's University, 923 Robie St., Halifax, Nova Scotia, B3H 3C3, Canada
e-mail: cgeroux@ap.smu.ca; bdeupree@ap.smu.ca

Geroux, C. and Deupree, R.G.: *2D Simulations of the Interaction Between Radial Pulsation and Convection*. Astrophys Space Sci Proc. **31**, 11–15 (2013)
DOI 10.1007/978-3-642-29630-7_2,

small perturbations resulting from machine round off. He was able to reproduce the location of the observed red edge of the RR Lyrae instability strip, finding that convection in the hydrogen ionization zone transports energy during contraction, while during expansion convection is not as effective at transporting energy, effectively removing the ability of the κ-mechanism to drive pulsation. Computers of 1977 were limited in their capabilities, and Deupree was only able to perform his calculations in 2D with limited resolution. Recently Gastine and Dintrans [4] have explored the interaction of convection and pulsation in 2D and have found that under the correct circumstances convection can quench pulsation. However, they used an ideal gas and an analytic opacity, so that comparison with observations is not possible. It is also widely accepted that 2D large scale turbulent flow patterns are different from 3D flow patterns.

Deupree's 1977 calculations were not Lagrangian, allowing the mesh to follow the pulsations by using a radial grid velocity that was the horizontal average of the radial velocity. This algorithm assumes that there is no net radial convective flow of material, which is not true, but the radial grid velocity does not impact the calculations other than redefining the zoning. Deupree's radial zoning drifted with respect to the structure and at later times the zoning was not fine enough in the H-ionization zone, and the calculations ran into numerical difficulties removing the ability to calculate full amplitude solutions. We have overcome these difficulties in our new radiation diffusion hydrodynamics code.

2 Radiation Diffusion Hydrodynamics Code

We have developed a new radiation diffusion hydrodynamics code, Stellar Pulsation with a Horizontal Eulerian, Radial Lagrangian Scheme (SPHERLS). It is capable of 1D fully radiative calculations or 2D and 3D calculations including convection. It has an Eulerian grid in the horizontal direction and uses M_r as the radial coordinate. Thus, the mass of radial shells remains constant. A grid velocity is chosen which does this; see Geroux and Deupree [6]. In 1D the use of this radial coordinate truly is Lagrangian, however in 2D and 3D it is only Lagrangian in a horizontally averaged sense, with material flowing freely between radial zones. It has been designed from the ground up for parallelism with MPI through domain decomposition. In the deep interior of the star SPHERLS uses a 1D region to avoid the increasingly small horizontal zoning that together with an increasingly large sound speed can dramatically reduce the time step. A realistic equation of state and opacities from Rogers and Iglesias [7, 10] and molecular opacities from Alexander [2] have been included. Near the surface the energy equation is solved implicitly to avoid the time-step being restricted by the speed of light instead of the speed of sound as zones become optically thin.

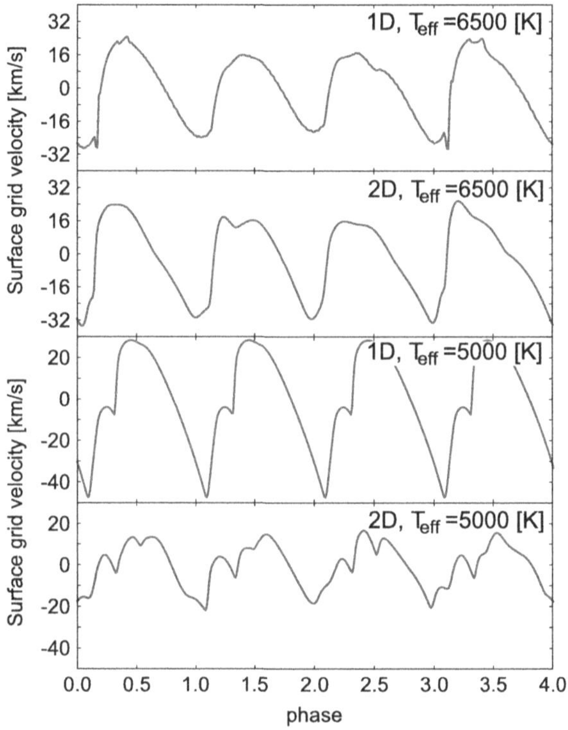

Fig. 1 The surface grid velocity, effectively the radial pulsation velocity, is plotted over four pulsation cycles for, from top to bottom, a 1D fully radiative model at 6,500 K, a 2D convective model at 6,500 K, a 1D fully radiative model at 5,000 K, and a 2D convective model at 6,500 K. The 5,000 K 2D convective model has a decrease in pulsation amplitude of almost 50% over the 1D fully radiative model. However, the 6,500 K models have nearly the same pulsation amplitude

3 Calculations and Results

Here we focus on the sizable array of 2D calculations we have completed. We have computed a range of ten angular-zone models with anywhere from 250 to 350 radial zones, with T_{eff} ranging from 6,500 to 5,000 K, helium mass fractions of 0.3 and 0.24, and angular spacings of 0.1°. We have also computed a few one hundred angular-zone calculations, at T_{eff} of 6,100 K and 5,700 K and helium mass fraction of 0.24, finding only small differences in the pulsation growth rates between the 10 and 100 angular-zone calculations. Two different angular spacings were testing, 0.1° and 0.03°. All models extend down to more than 96% of the stellar radius (to temperatures of more than 6×10^6 K) and all have luminosities of $50 L_{\odot}$. There are also 1D fully radiative counterparts for all multidimensional calculations.

Figure 1 shows comparison between surface velocity curves of 1D fully radiative and 2D convective calculations at 6,500 K and 5,000 K. The 1D and 2D models at 6,500 K, have very similar amplitudes, while at 5,000 K the surface velocity curve of the 2D convective model shows a nearly 50% decrease in amplitude from the 1D fully radiative model.

We have explored the dependence of the strength of convection on the phase of pulsation finding that during contraction both the radial, and theta convective velocities are a maximum, while just before full expansion the convective velocities

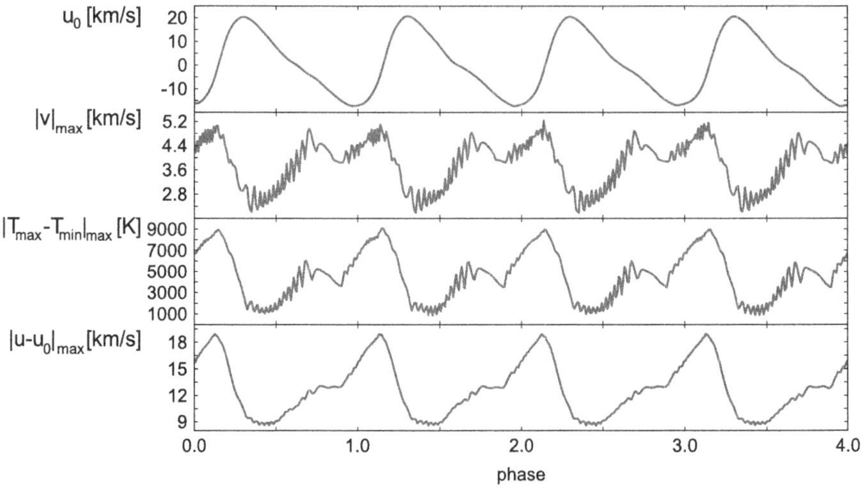

Fig. 2 Plotted as a function of pulsation phase from top to bottom are: the radial grid velocity at the surface of the model, the maximum absolute theta velocity, the maximum absolute temperature variation across horizontal zones over the entire model, and finally the maximum absolute radial convective velocity. These plots show the time dependence of convection, with larger convective velocities, and temperature variations during contraction, and smaller convective velocities and temperature variations during expansion

are at a minimum, as shown in Fig. 2. This time dependence is the same as that found by Deupree [3], where this time dependence leads to the quenching of the pulsation. Convection allows the flux through the hydrogen ionization zone during contraction, providing the outer layers of the star with an outward push; similarly during expansion the amount of flux transported by convection through the hydrogen ionization zone decreases the outward push, essentially negating the kappa effect.

As of yet, we have been unable to find a red edge to the instability strip over the temperature range of 6,500–5,000 K. While Deupree [3] was able to find a red edge, there are many numerical differences between SPHERLS and Deupree's 1977 code, nearly all of which are improvements. We are still examining the effects of the differences between the two codes on the identification of a red edge.

Acknowledgements CG would like to thank ACEnet for their graduate fellowship.

References

1. Bono, G., Caputo, F., Stellingwerf, R.F.: Pulsation and stability of RR Lyrae stars. I: Instability strip. ApJS **93**, 233–269 (1994)
2. Alexander, D.R., Ferguson, J.W.: Low-temperature Rosseland opacities. ApJ **437**, 879–891 (1994)
3. Deupree, R.G.: The theoretical red edge of the RR Lyrae gap. I – Dependence of convection on pulsation phase. ApJ **211**, 509–526 (1977)

4. Gastine, T., Dintrans, B.: Convective quenching of stellar pulsations. A&A **528**, A6 (2011)
5. Gehmeyr, M.: On nonlinear radial oscillations in convective RR Lyrae stars. I. – The mathematical description. ApJ **399**, 265–283 (1992)
6. Geroux, C.M., Deupree, R.G.: Radial stellar pulsation and three-dimensional convection. I – numerical methods and adiabatic test cases. ApJ **731**, 18–26 (2011)
7. Iglesias, C.A., Rogers, F.J.: Updated opal opacities. ApJ **464**, 943–953 (1996)
8. Marconi, M., Caputo, F., Di Criscienzo, M., Castellani, M.: RR Lyrae stars in galactic globular clusters. II. A theoretical approach to variables in M3. ApJ **596**, 299–313 (2003)
9. Marconi, M.: RR Lyrae pulsation theory. Am. Inst. Phys. Conf. Ser. **1170**, 223–234 (2009)
10. Rogers, F.J., Swenson, F.J., Iglesias, C.A.: OPAL equation-of-state tables for astrophysical applications. ApJ **456**, 902–908 (1996)
11. Stellingwerf, R.F.: Convection in pulsating stars. I – Nonlinear hydrodynamics. ApJ **262**, 330–343 (1982)

Vibrational Instability of Metal-Poor Low-Mass Main-Sequence Stars

Takafumi Sonoi and Hiromoto Shibahashi

Abstract We find that low-degree low-order g-modes become unstable in metal-poor low-mass stars due to the ε-mechanism of the pp-chain. Since the outer convection zone of these stars is limited only to the very outer layers, the uncertainty in the treatment of convection does not affect the result significantly. The decrease in metallicity leads to decrease in opacity and hence increase in luminosity of a star. This makes the star compact and results in decrease in the density contrast, which is favorable to the ε-mechanism instability. We find also instability for high order g-modes of metal-poor low-mass stars by the convective blocking mechanism. Since the effective temperature and the luminosity of metal-poor stars are significantly higher than those of Pop I stars, the stars showing γ Dor-type pulsation are substantially less massive than in the case of Pop I stars. We demonstrate that those modes are unstable for about 1 $M_{\odot}$ stars in the metal-poor case.

1 Introduction

The structure and evolution of stars born in metal-poor environments or when heavy elements were significantly deficient in the universe are considerably different than for stars born much later with higher heavy-element abundances. This is mainly because opacities and nuclear reactions are highly dependent on metallicity. Low metallicity leads to decreased opacity, and hence it makes luminosity of the star higher. On the other hand, the decrease in metallicity makes CNO-cycle energy generation less efficient. To maintain energy equilibrium in this situation, the stars with low metallicity need to be compact compared with Pop I stars. As a result,

T. Sonoi (✉) · H. Shibahashi
Department of Astronomy, University of Tokyo, 7-3-1 Hongo, Bunkyo-ku, Tokyo 113-0033, Japan
e-mail: sonoi@astron.s.u-tokyo.ac.jp; shibahashi@astron.s.u-tokyo.ac.jp

Sonoi, T. and Shibahashi, H.: *Vibrational Instability of Metal-Poor Low-Mass Main-Sequence Stars*. Astrophys Space Sci Proc. **31**, 17–21 (2013)
DOI 10.1007/978-3-642-29630-7_3,

the main sequence of metal deficient or metal poor stars on the HR diagram moves toward the bluer and higher luminosity side. The difference in structure must be conspicuous particularly for stars about 1.5 $M_\odot$, around which the main nuclear energy source changes between the CNO-cycle and the pp-chain and also the thickness of convective envelope drastically changes. Oscillation properties and stability of the stars are also expected to change significantly with metallicity. To confirm this, we perform a fully nonadiabatic analysis of metal-poor low-mass main-sequence stars.

2 ε-Mechanism for Low-Degree Low-Order g-Modes

In the pp-chain, at the stellar center ^{3}He is consumed by ^{3}He(^{3}He,2H)^{4}He or by ^{3}He(^{4}He,γ)^{7}Be immediately after it is generated. However, since the ^{3}He reactions are highly sensitive to temperature, they do not efficiently occur in the outer part of the nuclear burning region. As a consequence, ^{3}He accumulates in an off-centered shell. In such a situation, the high temperature dependence of the ^{3}He reactions might excite g-modes having a large amplitude in the off-centered ^{3}He shell. Such instability and the resultant material mixing was once proposed as a possible solution to the solar neutrino problem [1], and the stability of the sun was examined by several groups [2–5]. The presence of a convective envelope, which occupies the outer 20–30% in the case of Pop I low-mass stars, however, has made it hard to reach a definite conclusion on the stability. But the situation is different in metal-poor stars. As the metallicity decreases, the convective envelope becomes thinner and limited only to very outer layers (left panel of Fig. 1). In such a case, we can obtain a definite conclusion on the instability without the uncertainty due to the convective envelope. Indeed, the present authors showed that Pop III stars are unstable against low-order, dipole g-modes due to the ε-mechanism of the pp-chain [6, 7]. To extend our analyses to stars with $Z \neq 0$ but still having only very thin convective envelope, we examine the stability of stars with extremely low metallicity; $Z = 0.0005$ and $Z = 0.0001$ with the help of the stellar evolution code MESA [8] and a nonadiabatic pulsation code [7].

The right panel of Fig. 1 shows the results of our nonadiabatic analysis of those stars against the dipole ($l = 1$) g_1-mode. A wide range of models are found to become unstable due to the ε-mechanism. By comparing the stability analyses of stars with two different metallicities, we find that as the metallicity decreases the mass range of this instability extends toward higher mass. The reason for this is the tendency for the CNO-cycle to be replaced by the pp-chain for more massive stars grows as the metallicity and the CNO abundance decreases. The core of the metal-poor stars is convective at the ZAMS stage even with the pp-chain burning because of the high central temperature, and the gravity waves are evanescent there. But as the convective core shrinks and disappears with stellar evolution, the gravity waves can then propagate in the nuclear burning region. This is favorable for the ε-mechanism to work for instability.

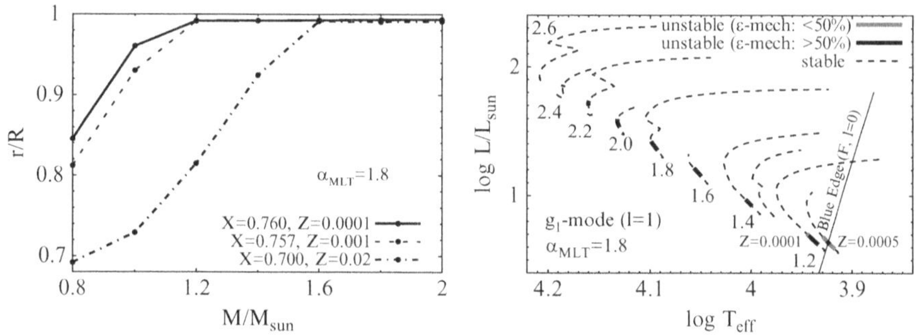

Fig. 1 *Left:* Location of base of convective envelope for ZAMS models with different compositions. *Right:* Evolutionary tracks on HR diagram for 1.2–2.6 $M_\odot$ with $X = 0.760$, $Z = 0.0001$ and 1.2–1.4 $M_\odot$ with $X = 0.759$, $Z = 0.0005$. The *dashed* and *thick solid lines* indicate the evolutionary stages at which the dipole ($l = 1$) g_1-mode is stable and unstable, respectively. The evolutionary stage in which the g_1-mode is unstable mainly due to the ε-mechanism is shown by *black thick lines*, while the κ-mechanism by *gray thick lines*

We find that the stars with $M \lesssim 1.2M_\odot$ become vibrationally unstable in the case of $Z = 0.0005$, and those with $M \lesssim 2.4M_\odot$ become unstable in the case of $Z = 0.0001$. The more massive stars keep the convective cores because of the dominant contribution of the CNO-cycle rather than the pp-chain. This is not favorable for instability, and low-degree low-order g-modes are not excited indeed in those stars. It should be remarked that in stars with $\log T_{\rm eff} \lesssim 3.95$, corresponding to the gray thick line parts of the evolutionary tracks in the right panel of Fig. 1, low-degree low-order g-modes are excited mainly by the κ-mechanism.

3 Low-Degree High-Order g-Modes (γ Doradus Type Oscillations)

Main-sequence stars near the cool end of the classical strip, called the γ Dor stars, show low-degree high-order g-mode pulsations. These stars have a thick convective envelope, and treatment of convection becomes an important issue in examining the stability analyses. Guzik et al. [9] noted the relatively long convective turn-over times at the base of the convective envelope of these stars. They claimed that a convective flux perturbation would be almost frozen-in and that excitation of the observed g-modes would be related with this characteristics and the fact that radiation becomes less responsible for energy transport from the base to the middle part of the convective envelope. Analyses with a time-dependent convection theory [10] demonstrated the above process. Based on these ideas, we adopt here a "frozen-in convection" approximation and simply ignore the convective flux perturbation to examine the stability of the γ Dor type oscillations of metal-poor stars. The instability region of high-order g-modes thus found in the metal-poor case is in the

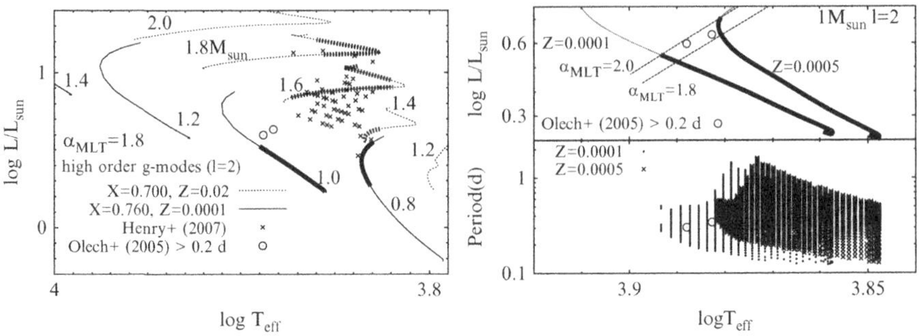

Fig. 2 *Left:* The *thick parts* of the evolutionary tracks correspond to unstable stages of high-order g-modes ($l = 2$). The *crosses* are the γ Dor stars reported by [12], while the *open circles* are stars, according to [11], having oscillation periods longer than 0.2 day. *Top of Right:* Evolutionary tracks with $Z = 0.0001$ and 0.0005. The *thick parts* show the evolutionary stage in which high-order g-modes are excited. The *dashed lines* are the blue edges of the instability region of the high order g-modes ($l = 2$) for $\alpha_{\rm MLT} = 1.8$ and 2.0, respectively. *Bottom of right:* The *crosses* and *dots* denote periods of the unstable modes ($l = 2$)

same temperature range as that in the Pop I case (Fig. 2). But, since the evolutionary tracks move toward bluer and higher luminosity side as metallicity decreases, the instability appears for the less massive stars compared with the Pop I case.

Olech et al. [11] found oscillations of blue stragglers in ω Cen, and detected many SX Phe stars, known as high amplitude δ Sct-like Pop II stars. They also detected oscillations having much lower frequencies than that of the radial fundamental mode. The corresponding stars are located around the blue edge of our theoretical instability region. Their periods correspond to the quadrupole ($l = 2$) modes. Although the possibility exists that such low-frequency oscillations might be tidally induced, they might be the first examples corresponding to the modes investigated in the present analysis.

Acknowledgements T. S. thanks the Hayakawa Satio Fund for its financial support.

References

1. Dilke, F.W.W., Gough, D.O.: Nature **240**, 262 (1972)
2. Dziembowski, W., Sienkiewicz, R.: Acta Astronom. **23**, 273 (1973)
3. Christensen-Dalsgaard, J., Dilke, F.W.W., Gough, D.O.: MNRAS **169**, 429 (1974)
4. Boury, A., Gabriel, A., Noels, A., Scuflaire, R., Ledoux, P.: A&A **41**, 279 (1975)
5. Shibahashi, H., Osaki, Y., Unno, W.: PASJ **27**, 401 (1975)
6. Sonoi, T., Shibahashi, H.: PASJ **63**, 95 (2011)
7. Sonoi, T., Shibahashi, H.: PASJ **64**, 2 (2012)
8. Paxton, B., Bildsten, L., Dotter, A., Herwig, F., Lesaffre, P., Timmes, F.: ApJS **192**, 3 (2011)

9. Guzik, J.A., Kaye, A.B., Bradley, P.A., Cox, A.N., Neuforge, C.: ApJ **542**, L57 (2000)
10. Dupret, M.A., Grigahcéne, A., Garrido, R., Gabriel, M., Scuflaire, R.: A&A **414**, L17 (2004)
11. Olech, A., Dziembowski, W.A., Pamyatnykh, A.A., Kaluzny, J., Pych, W., Schwarzenberg-Czerny, A., Thompson, I.B.: MNRAS **363**, 40 (2005)
12. Henry, G.W., Fekel, F.C., Henry, S.M.: AJ **133**, 1421 (2007)

Non-radial Pulsation and Magnetic Fields of OB Stars

Natallia Sudnik and Alexander Kholtygin

Abstract We report the results of our studying the fast line-profile variability (LPV) in spectra of bright OB stars. Spectral and spectropolarimetric observations of 15 OB stars were made during the last ten years in the framework of the program of searching for the regular and stochastic LPV in spectra of OB stars and their magnetic fields. The spectra were obtained using the 6-m telescope of the Special Astronomical Observatory (SAO, Russia) and the 1.8-m telescope of the Bohyunsan Optical Astronomy Observatory (BOAO, Korea). For all studied stars we detected regular LPV connected with non-radial pulsation (NRP) and induced by the rotation of the large-scale structures in the stellar wind. We propose that the NRP are a trigger of the formation of the large-scale structures in the stellar wind. Results of searching for the polarimetric line profile variability (pLPV) in the spectra of OB stars are also reported. We find that the pattern of LPV can differ for the left and right polarized components of stellar radiation.

1 Observations

The observations analyzed were made in 1997–2010 at the SAO and BOAO. More than 1000 spectra of 15 OB stars (ξ Per, ϵ Per, α Cam, HD 35502, δ Ori, λ Ori A, θ^1 Ori C, ζ Ori A, 15 Mon, ρ Leo, HD 93521, 85 Her, 19 Cep, λ Cep and 10 Lac) were obtained. The 6-m telescope observations were made with the Lynx spectrograph (spectral resolution $R = 60,000$) and a CCD with 512×512 pixels, with the NES spectrograph ($R = 60,000$) and a 1×1k CCD and with the MSS spectrograph ($R = 15,000$) and a 2k $\times$ 2k CCD, while the 1-m telescope observations were made with the CEGS spectrograph ($R = 45,000$) and a $1,242 \times 1,152$ CCD. Most of these

N. Sudnik (✉) · A. Kholtygin
Saint-Petersburg State University, Universitetskii 28, Saint-Petersburg, 198504, Russia
e-mail: snata.astro@gmail.com; afkholtygin@gmail.com

Sudnik, N. and Kholtygin, A.: *Non-radial Pulsation and Magnetic Fields of OB Stars.*
Astrophys Space Sci Proc. **31**, 23–26 (2013)
DOI 10.1007/978-3-642-29630-7_4,

stars were also observed in BOAO by using the 1.8-m telescope equipped with the BOES spectrograph ($R = 45,000$) and large 2×4k CCD. The reduction of SAO spectra was made with MIDAS while the BOAO spectra were processed with IRAF.

2 Line Profile Variability

The amplitudes of the line profile variations in the spectra of O and early B stars are small, so we call this line profile variability as microvariability (microLPV). To detect microLPV we use smoothing Time Variation Spectrum (smTVS) analysis introduced by Kholtygin et al. [2]. In this method before the standard deviation spectrum was obtained, the differential spectra were smoothed using a wide Gauss filter (S). After smoothing the amplitude of a noise component of the differential profiles decreases by a factor of $\approx \sqrt{S/\delta\lambda}$, where $\delta\lambda$ is a pixel width. If the width of the variable component is not smaller than that of the filter, then smoothing will not significantly change the amplitude of the variable component, and a peak in the standard deviation spectrum that corresponds to the variable component can be detected. The smTVS is described by

$$smTVS(\lambda, S) = \frac{1}{N-1}\left(\sum_{i=1}^{n} g_i \left[I_i(\lambda, S) - \overline{I(\lambda, S)}\right]^2 \Big/ \sum_{i=1}^{n} g_i\right), \qquad (1)$$

where N is the number of spectra, $I_i(\lambda, S)$ is the line intensity in the spectrum number i at wavelength λ smoothed with Gaussian filter (S is a filter width) and normalized to the continuum level, $\overline{I(\lambda, S)}$ is the mean intensity at wavelength λ averaged over all smoothed line profiles, g_i is a relative weight of the ith observation. The smTVS analysis is a very sensitive tool to detect micro-variations of line profiles. It can be used when the amplitude of variations is small and do not exceed the noise level, when the number of spectra is small and the time grid is uneven.

For $S = 0$ the value of smTVS(λ,0) corresponds to the TVS (Temporal Variance Spectrum) value introduced by Fullerton et al. [1]. The best results are obtained for smoothing with a Gaussian filter with S =0.7–1.3 Å (15–30 pixels).

In Fig. 1 (left panel) we present the density plot of the smTVS for the region of the H_γ line in spectra of λ Ori. Darker areas correspond to higher amplitudes of smTVS. Clearly the variations of the H_γ line profiles can be seen at all filter widths. The smTVS with the filter width more than 1 Å also indicates variability of the weak C II-III N II, O II and Si IV line profiles. In a Fig. 1 (right panel) TVS and smTVS functions are shown for comparison. The variations of weak lines can be easily detected using the smTVS analysis.

To search for the regular LPV we carried out Fourier analysis using the CLEAN procedure [3] modified in [5]. LPV periods detected by us for the program stars are given in Table 1. Both long periods of LPV connected with rotational modulation

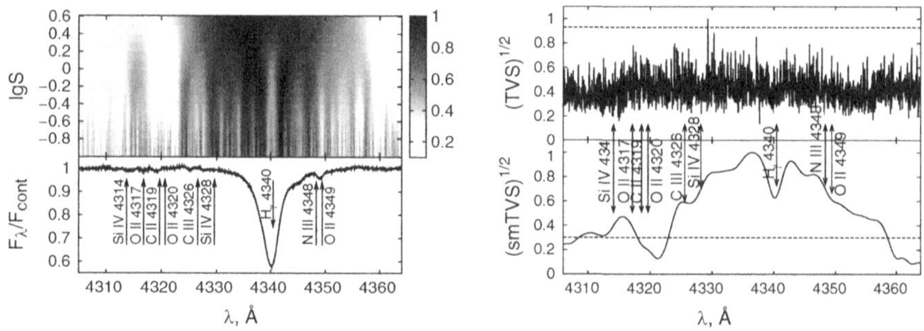

Fig. 1 *Left panel:* smTVS map (*top*) and mean line profile (*bottom*) in the region λλ4308–4362 Å in spectra of λ Ori. *Right panel:* TVS (*top*) and smTVS (*bottom*) for lines in the same interval as in the left panel. Filter width is 1.26 Å. The dashed line marks the significance level $\alpha = 0.001$

Table 1 Period detected in the program stars LPVs

Name	Spectral type[a]	Period	Name	Spectral type[a]	Period
λ Cep	O6.5 I f(n)p	$3–11^h$, $0.6–3.4^d$	ζ Ori A	O9 Ib	$3–6^h$
λ Ori A	O8 III ((f))	$3–16^h$, $1.3–1.8^d$	ι Her	B3 IV	$7–13^h$, $0.9–2.9^d$
δ Ori A	O9.5 II	4.1^h	ρ Leo	B1 Iab	$3–6^h$, $0.6–1.8^d$

[a] from [4] and from SIMBAD data base for ι Her and ρ Leo

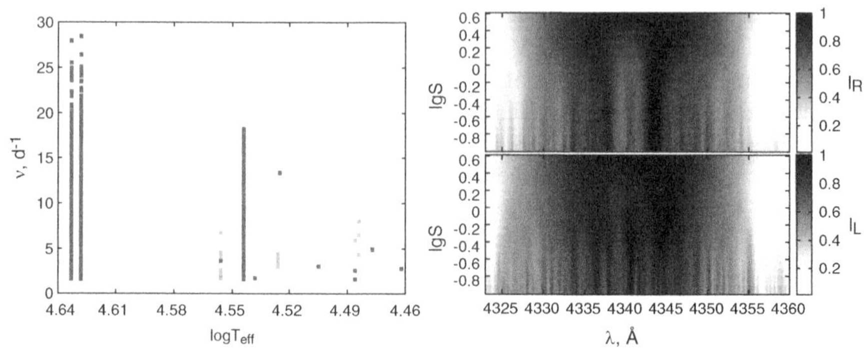

Fig. 2 *Left panel:* detected NRP frequencies for O stars as function of the effective temperature of the stars. *Right panel:* Density plot diagram of smTVS for I_R (*top*) and I_L (*bottom*) components of line profiles in spectra of ζ Ori

(≈$0.6–3.5^d$), and short periods attributed to NRP (≈$3–8^h$) were found. We plot in Fig. 2 (left panel) the frequencies of NRP for O stars taken from the literature and including our own data. More than ten well-studied O stars seem to be non-radial pulsators. We suppose that NRP is the common feature of all hot massive stars.

3 Polarimetric Line Profile Variability

We study LPV in integral (Stokes parameter I), left (I_L) and right (I_R) polarized components of the line profile in spectra of stars λ Cep, ζ Ori A and λ Ori A. We find that pattern of LPV for I_L and I_R can differ.

Applying Eq. (1) to the left-polarized components I_L^i of the line we obtain the smTVS$_L$ spectra and the smTVS$_R$ spectra in the case of components I_R^i. The density plot of the smTVS$_L$ and smTVS$_R$ spectra for the region near H$_\gamma$ line in spectra of the star ζ Ori is shown in the Fig. 2. Darker areas correspond to higher amplitudes of the smTVS. The differences between smTVS$_L$ and smTVS$_R$ spectra are clearly seen and can be indirect evidence that the stars possess a magnetic field.

We also compare the Fourier spectra of the I_L and I_R components of the lines in the spectra of the program stars. We find that there are regular components which can be detected in the Fourier spectrum for the I_L component of line and cannot be detected for the I_R component and vice versa. We do not know the cause of this difference. Possibly it is connected with the clumps in the stellar wind and the rugged structure of the magnetic field above the photosphere.

4 Conclusions

All studied stars show microLPV with the amplitude about 0.5–3%. Regular short and long timescale components of LPV have been detected. These components are probably connected with non-radial pulsations and rotation modulation. NRP seems to be common for all massive OB stars.

Weak variability of lines of the ions Si, C, O, N in spectra of the stars λ Cep, λ Ori A, δ Ori A and ζ Ori A was revealed by smTVS. It is possible that the LPV for the left and right-polarized components of the line profiles can differ.

Acknowledgements This study is supported by the Project 6.38.73.2011 of the Saint-Petersburg State University (SPbSU).

References

1. Fullerton, A.W., Gies, D.R., Bolton, C.T.: J. Suppl. Ser. **103**, 475–512 (1996)
2. Kholtygin, A.F., Monin, D.N., Surkov, A.E., Fabrika, S.N.: AstL. **29**(3), 175–187 (2003)
3. Roberts, D.H., Lehar, J., Dreher, J.W.: Astrophys. J. **93**(4), 968–989 (1987)
4. Sota, A., Maíz Apellániz, J., Walborn, N.L., et al.: Astrophys. J. Suppl. Ser. **193**, 24–74 (2011)
5. Vityazev, V.V.: Analysis of Irregular Time Series. SPbSU, Saint-Petersburg (2001)

Pulsations in Hot Massive Stars

C.C. Lovekin and J.A. Guzik

Abstract Massive stars are known to pulsate at many stages of evolution. Most spectacular are the Luminous Blue Variables (LBVs), in which pulsation is one possible origin for the S-Dor type outbursts. In this work, we study the radial pulsations of stars with initial masses of 20, 40, 60 and 85 $M_{\odot}$ using both linear and non-linear pulsation codes. The pulsations can then interact with time-dependent convection, which increases the luminosity until the Eddington limit is exceeded locally, potentially driving mass loss and S-Dor outbursts. We consider models at various stages of evolution and metallicity, covering the observed properties of the majority of the observed LBV and LBV candidates. Preliminary results characterizing the pulsations as functions of Y and Z are presented.

1 Introduction

S Dor stars are massive stars which undergo periods of high mass loss. Although these outbursts are much less dramatic than the giant eruptions seen in some luminous blue variables (LBVs) (e.g., η Car), the amount of mass lost can still be significant. S Dor outbursts occur on time scales of months to years, and typically eject around 10^{-4} $M_{\odot}$ per event. During an S Dor outburst, the visual magnitude increases by 1–2 magnitudes, while the bolometric magnitude of the star remains approximately constant (see, eg., Vink [9] for a review).

C.C. Lovekin (✉)
Los Alamos National Laboratory, P.O. Box 1663, T-2, MS-B283, Los Alamos, NM 87545, USA
e-mail: clovekin@lanl.gov

J.A. Guzik
Los Alamos National Laboratory, P.O. Box 1663, XTD-2, MS-T086, Los Alamos, NM 87545, USA
e-mail: joy@lanl.gov

Lovekin, C.C. and Guzik, J.A.: *Pulsations in Hot Massive Star*.
Astrophys Space Sci Proc. **31**, 27–31 (2013)
DOI 10.1007/978-3-642-29630-7_5,

The origin of the variability of these objects is not clear. It was initially proposed that increased mass loss caused the star to form a pseudo-photosphere, resulting in an apparent decrease in temperature. This has now been shown to be incorrect, and the variations in stellar properties are a result of changes in the stellar radius [2, 6]. The root cause of these variations in radius are still not understood.

In this work, we continue previous investigations into pulsations as the possible origin of S Dor outbursts. It has been suggested that κ-effect pulsations arising from the iron bump near 200,000 K, combined with time-dependent convection could produce a super-Eddington layer, causing the star to become unstable and drive mass loss [3–5, 8]. We extend this study to lower masses, and use more detailed stellar models, based on evolutionary tracks calculated by Meynet et al. [7].

2 Models

We use published evolutionary tracks [7] to calculate models of pulsating stars with masses from 20–85 $M_{\odot}$ and metallicites from $Z = 0.004$–0.04. We calculate pulsation frequencies using a linear non-adiabatic (LNA) code in envelope models, including the outer 3–7% of the stellar mass. The mass of the envelope model is adjusted to ensure that the temperature at the base is at least 2 million Kelvin, so that the Fe ionization region is included. Once the frequencies have been calculated, the envelope models are put into a 1D hydrodynamic code (DYNSTAR) [1], and given an initial velocity of 1 km s^{-1} in the chosen mode (fundamental, first overtone, etc.) The hydrodynamic models, which include a prescription for time-dependent convection, were then simulated for up to a few hundred days.

3 Results

We found models unstable to pulsation in a wide variety of models across the full range of our parameter space. Based on these results, we can consider the effects of mass and metallicity on the pulsation frequencies and amplitudes. Although we do not explicitly change the He abundance of our models, radiatively driven mass loss removes the outer layers of the star in the more massive and metal-rich evolutionary models, so increasing age can be used as a proxy for increased He abundance. We analyze this effect in the case of a 60 $M_{\odot}$ model at $Z = 0.04$. The surface radial velocities are shown at three different ages for this model in Fig. 1. As the star ages (from top to bottom) the He abundance also increases (from $Y \sim 0.3$ to ~ 0.9). At first glance, increased He abundance increases the period and decreases the amplitude. However, the mass, luminosity, radius and temperature all change as the model ages as well. We use the mean density to clarify the situation. If we assume that $Q (= \Pi(\overline{\rho}/\overline{\rho_{\odot}})^{1/2})$ should be constant and all three models are pulsating in the fundamental mode, we can use the Q of the first model to predict

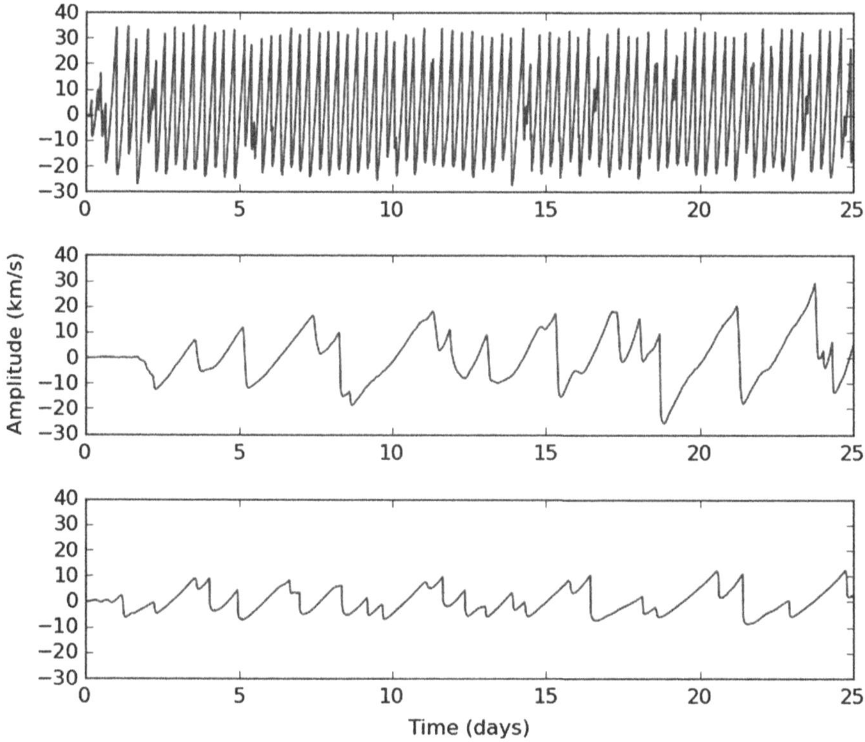

Fig. 1 The radial velocity curves for a $60M_{\odot}$ model with $Z = 0.04$. The surface helium abundance, from *top to bottom*, is $Y = 0.34$, $Y = 0.5266$, and $Y = 0.9536$

the periods of the other two models. In this case, the models are all close to the same evolutionary state, so their overall structure should be similar. As a result, we expect the assumption of constant Q to be reasonable. We find that the periods for the middle and bottom model are longer than expected. If the first model is actually pulsating in the first harmonic, the discrepancy is decreased, but not eliminated. We feel confident in concluding that increased He abundance results in longer pulsation periods.

3.1 Outbursts

A few models considered here show outburst-like behavior, with the surface radial velocity increasing to 60–80 km s^{-1}. In stars this massive, these velocities are still well below the escape velocity. However, the sudden increase in surface radius, and corresponding increase in luminosity to a few times the Eddington luminosity will increase mass loss rates. In at least one model, the surface luminosity becomes

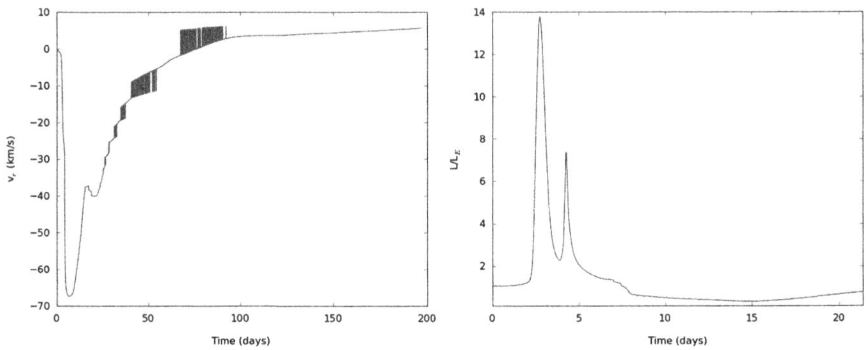

Fig. 2 The surface radial velocity (*left*) and surface luminosity as a fraction of the Eddington luminosity (*right*) for an 85 $M_\odot$ model with $Z = 0.004$. This model falls close to the location of other known S Dor variables

greater than the Eddington luminosity for a period of a few days. The surface radial velocity for one of these models is shown in Fig. 2, along with the corresponding luminosity as a fraction of the Eddington luminosity. The period of rapid expansion corresponds to a period of super-Eddington luminosity, and lasts a few days.

4 Conclusions

In the models considered here, the pulsation amplitudes can become quite large, particularly as the mass of the model increases. A few of the models, particularly at 85 $M_\odot$, show outburst-like events. The surface radial velocities remain below the escape velocity of the model, so it seems that radial pulsations alone will not drive significant mass loss. However, the increase in surface radius and luminosity may lead to increased mass loss rates, which could explain the S Dor variability.

Acknowledgements This work was performed for the U. S. Department of Energy by Los Alamos National Laboratory under Contract No. DE-AC52-06NA2-5396.

References

1. Cox, A.N., Ostlie, D.A.: Ap&SS **210**, 311 (1993)
2. de Koter, A., Lamers, H.J.G.L.M., Schmutz, W.: A&A **306**, 501 (1996)
3. Guzik, J.A., Cox, A.N., Despain, K.M.: The Fate of the Most Massive Stars, vol. 332, p. 263. ASP, San Francisco (2005)
4. Guzik, J.A., Cox, A.N., Despain, K.M., Soukup, M.S.: IAU Colloq. 169: Variable and Non-spherical Stellar Winds in Luminous Hot Stars, vol. 523, p. 337 (1999)
5. Guzik, J.A., Cox, A.N., Despain, K.M.: Eta Carinae at The Millennium, vol. 179, p. 347 (1999)

6. Leitherer, C., Schmutz, W., Abbott, D.C., Hamann, W.-R., Wessolowski, U.: ApJ **346**, 919 (1989)
7. Meynet, G., Maeder, A., Schaller, G., Schaerer, D., Charbonnel, C.: A&AS **103**, 97 (1994)
8. Onifer, A.J., Guzik, J.A.: In: Bresolin, F., Crowther, P.A., Puls, J. (eds.) IAU Symposium, vol. 250, 83.0805.0764 (2008)
9. Vink, J.S.: In: Davidson, K., Humphreys, R. (eds.) Eta Carinae and the Supernova Imposters, (2012). arXiv:astro-ph/0905.3338

On the Origin of the Dense Frequency Spectra of Space Observed Intermediate Mass Pulsating Stars

J. Pascual-Granado, A. Grigahcène, D. Díaz-Fraile, M. Gruberbauer, R. Garrido, P. Amado and J. C. Suárez

Abstract Several hypotheses have been advanced to explain the wealthy spectra of space observed intermediate mass pulsating stars. Among which we can find the rotational splitting, stochastic excitation or detection of coloured noise due to superficial granulation. An overview of the problem requires a coherent analysis of a large number of stars looking for correlations between the observed frequency spectra and the position on the HR diagram.

1 Introduction

Recently some authors have claimed the detection of hundreds (HD 174936 in [3]) or even more than a thousand (HD 50844 in [10]) pulsational frequencies in observations of δ Scuti stars from the CoRoT satellite. This is far beyond any theoretical prediction.

It has been proposed that the multitude of frequencies detected is caused partly by non-white stellar background noise due to surface granulation [6].

The aim of this work is to give some insight into this question. Our objective is to check the possible relation between surface physical parameters of a sample of stars with the distribution of peaks observed in their frequency spectrum. We further

J. Pascual-Granado (✉) · D. Díaz-Fraile · R. Garrido · P. Amado · J. C. Suárez
Instituto de Astrofísica de Andalucía (CSIC), Glorieta de la Astronomía s/n 18008, Granada, Spain
e-mail: javier@iaa.es

A. Grigahcène
Centro de Astrofísica, Faculdade de Ciencias, Universidade do Porto, Rua das Estrelas, 4150-762 Porto, Portugal
e-mail: Ahmed.Grigahcene@astro.up.pt

M. Gruberbauer
Department of Astronomy & Physics, Saint Marys University, Halifax, NS B3H 3C3, Canada

Pascual-Granado, J. et al.: *On the Origin of the Dense Frequency Spectra of Space Observed Intermediate Mass Pulsating Stars*. Astrophys Space Sci Proc. **31**, 33–37 (2013)
DOI 10.1007/978-3-642-29630-7_6,

present some revealing HR diagrams comparing theoretical results about convection with the peak distribution.

2 Data Analysis

The full sample analyzed consists of 40 stars labelled as A/F stars in the CoRoT database selected independently of the variability class.

The light curves were detrended using polynomial fitting or a piecewise linear spline. For the correct detection of frequencies it is very important to remove all the undesired effects produced by the spectral window. Thus, a new gap-filling method presented recently [9] was used in order to interpolate the gaps in the data produced by the South Atlantic Anomaly. This method is based on ARMA models.

Frequencies were calculated using a peak detection algorithm based on a Scargle periodogram and assuming only white noise. We applied a Nuttall window function to remove the sidelobes of the peaks. We used a conservative limit for detection of Signal-Noise-Ratio: SNR>6. The error estimation of the detection was made using Montgomery and O'Donoghue approximation [8]. We investigated in all cases the range 0–1,200 μHz (approximately 0–100 c/d) which is the range where mesogranulation and granulation are more effective [7].

3 Physical Parameters

We gathered all data available about $T_{\rm eff}$, surface gravity, metallicities and absolute magnitudes from Corotsky database [1]. We searched also in the literature [2] to get values for some stars that were not present in those databases and to obtain additional, more recent estimates from updated catalogs. For F type stars we searched in the Geneva-Copenhagen survey of the solar neighbourhood [5]. For the stars for which we did not find any determination of $T_{\rm eff}$ or surface gravity we used Strmgren photometry and calculated physical parameters with TempLogg [12] and different calibrations. For this part we took the photometric data from GAUDI and the Hauck and Mermilliod catalog [4]. We tested also our determinations deriving for some stars the physical parameters from ground-based spectroscopic data and comparing the results.

4 HR Diagram

Figure 1 shows the HRD of all stars analyzed (black circles). The size of the circles is scaled with the number of peaks detected in every spectrum. The band represented is the instability strip and the coloured contours represent the number of modes excited of low degree ($\ell = 0 - 3$) as calculated using our theoretical models with solar metallicity and $\alpha = 1.8$. As can be seen, there is no coincidence between the

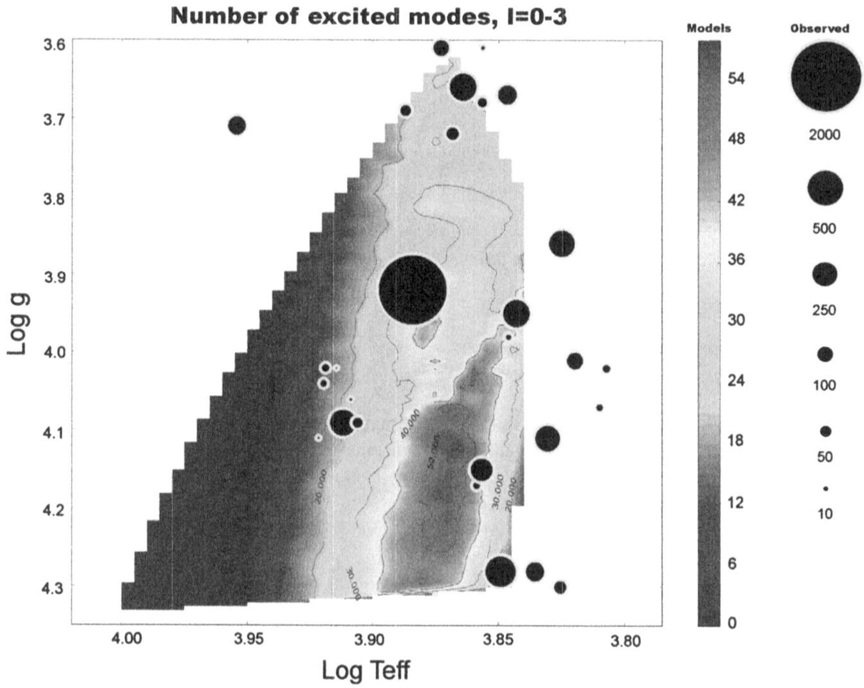

Fig. 1 Position of analyzed stars on HRD (*circles*). The size of the circles is scaled with the number of detected frequencies. Background colors represent the number of modes excited of low degree (ℓ =0–3) calculated using our theoretical models

isolines of the number of excited modes and the size of the circles. All this points that there is no direct connection between the number of modes observed and the number of unstable modes.

Left panel of Fig. 2 shows how the convective luminosity of the external layers changes during three stages of a star of 1.9 $M_{\odot}$ evolving from the blue edge (A), center (B) and red edge (C) of the instability strip, respectively. These three stages are marked on the right panel (red points) of Fig. 2 in the corresponding evolutionary track of the HR diagram. In the same panel four stars of the sample analyzed are overimposed as black circles showing the same scale for the relation between the size of the circles and the number of modes as in Fig. 1.

Stellar convection is still challenging and the relation between the convective luminosity and the granulation intensity is not known, but the low luminosity and the small depth shown at stages A and B impose a limit on the energy available.

Moreover, from both panels it can be deduced that no obvious correlation between the luminosity of the convective layer and the number of peaks detected in the frequency spectrum is found. In the light of these results any hypothesis about the contribution of granulation appears weak from a theoretical perspective.

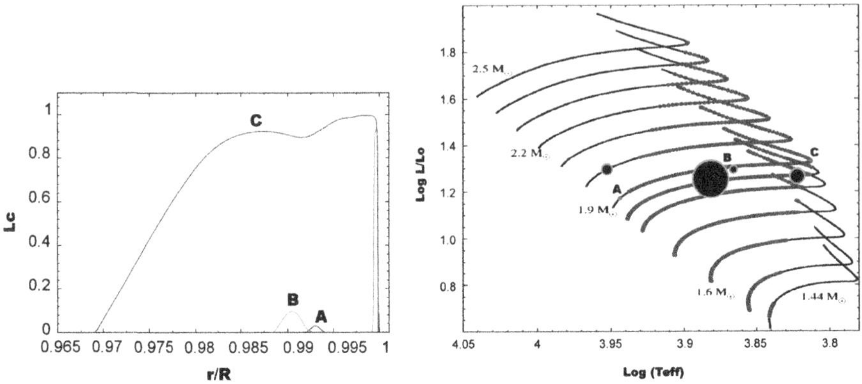

Fig. 2 *Left panel*: Convective luminosity of the external layers for the same star in three different evolutionary stages. *Right panel*: HRD showing evolutionary tracks (*black lines*) and three stars of the sample analyzed (*circles*). Points A, B, C corresponds to the stages represented in left panel. *Blue lines* corresponds to the instability strip

5 Conclusions

It has been suggested before that most of the peaks detected in the overcrowded spectrum of some delta Scuti stars are due to granulation noise. In another contribution [11] this hypothesis appears more controversial when HD 50844 is compared with a constant star (HD 171856) that has the same spectral type (A2) but does not show any trace of noise due to granulation. We evaluated this possibility exploring the hypothesis with theoretical models and a larger sample of stars. We showed that our models fail to reproduce enough pulsation modes to explain the observations. Furthermore, according to our models the convective properties of the observed stars are probably insufficient to produce significant granulation. Before attributing or not to the granulation any background signal, more critical work is needed in this direction. At the same time, this work could help us to better understand the granulation in solar-like stars hotter than the Sun like HD 49933 (F spectral type).

The question of the nature of the frequency spectrum of some intermediate mass pulsating stars is still open.

References

1. Charpinet, S., Cuvilo, J., Platzer, J., et al.: in ESA SP-1306, p. 353 (2006)
2. Prugniel, Ph., Soubiran, C., Koleva, M., Le Borgne, D.: arXiv:astro-ph/0703658
3. Garca Hernndez, A., Moya, A., Michel, E., et al.: A&A **506**, 79–83 (2009)
4. Hauck, B., Mermilliod, M.: A&A Suppl. **129**, 431–433 (1998)
5. Holmberg, J., Nordstrm, B., Andersen, J.: A&A **501**, 941–947 (2009)

6. Kallinger, T., Matthews, J.: ApJ **711**, L35–L39 (2010)
7. Michel, E. Samadi, R., Baudin, F., et al.: A&A **495**, 979–987 (2009)
8. Montgomery, M., O'Donoghue, D.: Delta Scuti Star Newslett. 13 (1999)
9. Pascual Granado, J., Garrido, R.: IAU Symp. **285**, 392 (2012)
10. Poretti, E., Michel, E., Garrido, et al.: A&A **506**, 85–93 (2009)
11. Poretti, E., Mantegazza, L., Niemczura, E., et al.: AN **331**, 9/10, p.1049 (2010)
12. Rogers, N.Y.: Comm. Asteroseismol. 78 (1995)

The Spectroscopic Observations of CoRoT Asteroseismic Targets with HARPS

Ennio Poretti, Monica Rainer, Luciano Mantegazza, Michele Floquet, Philippe Mathias, Pedro Amado, Conny Aerts, Katrien Uytterhoeven, and the CoRoT Seismology Ground-Based Working Group

Abstract CoRoT photometric measurements of asteroseismic targets need complementary ground-based spectroscopic observations. We are using the planet-hunter HARPS spectrograph attached to the 3.6m-ESO telescope in the framework of two consecutive Large Programmes. We discuss its use to study line-profile variations and we report on a specific result obtained for the δ Sct star HD 170699.

1 Introduction

Ground-based spectroscopic observations began at once with the launch of the CoRoT satellite. They involve several high-resolution échelle spectrographs: FEROS

E. Poretti (✉) · M. Rainer · L. Mantegazza
INAF-OA Brera, Via E. Bianchi 46, 23879 Merate, Italy
e-mail: ennio.poretti@brera.inaf.it; monica.rainer@brera.inaf.it; luciano.mantegazza@brera.inaf.it

M. Floquet
GEPI, Observatoire de Paris, 5 place Jules Janssen, 92195 Meudon Cedex, France
e-mail: michele.floquet@obspm.fr

P. Mathias
IRAP, 57 Avenue d'Azereix, 65008 Tarbes, France
e-mail: philippe.mathias@irap.omp.eu

P. Amado
Instituto de Astrofísica de Andalucía, Apartado 3004, 18080 Granada, Spain
e-mail: pja.iaa.es

C. Aerts
Instituut voor Sterrenkunde, KU Leuven, Celestijnenlaan 200D, 3001, Leuven, Belgium
e-mail: conny@ster.kuleuven.ac.be

K. Uytterhoeven
Instituto de Astrofísica de Canarias (IAC), Tenerife, Spain
e-mail: katrien@iac.es

Poretti, E. et al.: *The Spectroscopic Observations of CoRoT Asteroseismic Targets with HARPS*. Astrophys Space Sci Proc. **31**, 39–42 (2013)
DOI 10.1007/978-3-642-29630-7_7, © Springer-Verlag Berlin Heidelberg 2013

and HARPS at ESO-La Silla (Chile), FOCES at Calar Alto (Spain), SOPHIE at the Observatoire Haute Provence (France), FIES at NOT and HERMES at Mercator [6] (both at the Observatorio Roque de los Muchachos, La Palma, Spain). A small number of spectra were also obtained with CORALIE at ESO-La Silla and HERCULES at Mt. John Observatory (New Zealand). After the completion of the ESO Large Programme with FEROS (LP 178-D.0361, 60 nights in total), we obtained two subsequent Large Programmes with HARPS, the famous planet–hunter échelle spectrograph primarily used for the measurements of radial velocities. The first Large Programme with HARPS (LP 182.D-0356) consisted of 45 nights of observations between December 2008 and December 2009, while the current LP 185.D-0056 started on June 2010 and will end in January 2013. As of now, we have taken about 2500 HARPS spectra of CoRoT asteroseimic targets. The high-resolution spectra make the detection of high-degree modes possible [4] since the stellar disk can be spatially resolved thanks to the Doppler shifts induced by the star's rotation. Therefore, by measuring the observed variations in the line profiles, it is possible to know what kind of modes are excited in the stars and to assign the spherical wavenumbers (ℓ, m) to each of them. The HARPS spectra are reduced and normalized in a homogeneous way using a semi-automated pipeline developed at INAF-OAB.

2 HARPS: Improvements in the Instrument Setup

HARPS is able to operate in two configurations: high accuracy (HAM mode, with R = 115,000) or high efficiency (EGGS mode, with R = 80,000). At the telescope, the HARPS pipeline provides an estimate of the signal-to-noise ratio (SNR) at three different wavelengths (4,500, 5,000, and 6,500 Å). We chose to work mostly in the EGGS configuration, aiming at a signal-to-noise ratio (SNR) estimated at the telescope of about 200. In the reduction process we computed the SNRs taking into account photon noise, readout noise, and flat field correction. We could infer that for our stars (B-A-F spectral types) the SNR values given by the HARPS pipeline at 5,500 Å are a little too optimistic when compared with the values we measured between 5,802 and 5,825 Å. The ratio between the two SNR values is around 1.23 for A-F stars and around 1.28 for B stars.

Analysing the HARPS data obtained during the first observing runs (14–24 December 2008 and 3–8 January 2009) of the Large Programme ESO LP 182.D-0356, we found the presence of a disturbing feature in the spectra, i.e., an oscillation with an amplitude of about 0.05% of the continuum. It was the first time that this problem was reported. In fact, our approach to the data is different than the one usually used with the HARPS spectra. Usually the spectra are used to compute radial velocities of cool stars, while we search for line profile variations in hot stars. In particular, this feature became evident only by averaging a large number of spectra of the same line (Fig. 1, upper red curve), while it is practically invisible in a single spectrum. During the observing run of June 2009, several attempts were made to

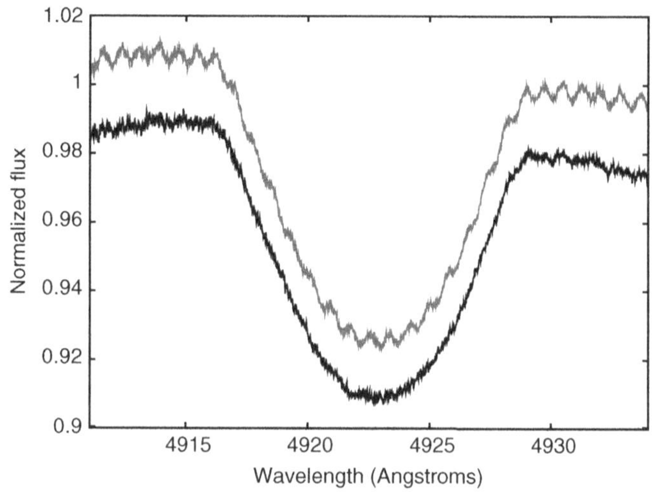

Fig. 1 Average 4,921 Å He I line of the Be star HD 51452: the instrumental effect (*upper red curve*) was eliminated by removing the blue filter located on the optical path of the HARPS calibration lamp (*lower black curve*)

understand the nature of this problem, which in the meantime had worsened. It was found that the filters in front of the calibration lamps were misaligned after a check-up in May 2009, and this enhanced the amplitude of the oscillations, which reached a value of about 2% of the continuum. Realigning the filters decreased the feature's amplitude to the previous 0.05% of the continuum, but did not eliminate it. With the help of the ESO staff we were able to identify the source of this problem as an interference created by the flat-field filters (red and blue). In fact, the issue was solved by taking away the blue filter: the subsequent flat-field correction was free from the instrumental effect (Fig. 1, lower black curve).

3 Evidence of High-Degree Modes Excited in δ Sct Stars

Several papers were published using the spectra observed in the ground-based programme (see the special A&A vol. 506) and a lot of work is still in progress. The data have been used in several ways, ranging from frequency analysis to the identification of the physical parameters of the stars. We present here the case of the δ Sct stars, which are main-sequence or slightly evolved stars showing multiperiodic light curves. CoRoT introduced a dramatic change in terms of both number of identified frequencies and of the involved physics: hundreds of frequencies were detected in the light curve of HD 50844 [5]. The main goal of the spectroscopic observations is to determine if the observed frequencies can be associated with pulsation modes. The mode identification was performed by fitting the amplitude

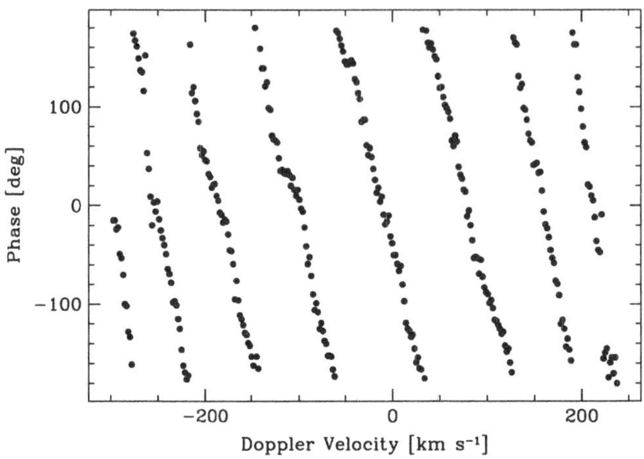

Fig. 2 HD 170699: phase variations across the line profile of the $f = 48.26\ \mathrm{d}^{-1}$ term (HARPS spectra)

and phase variations of each mode across the average line profile of each HARPS spectrum [3, 8]. Figure 2 shows the phase diagram of the high-frequency term $f = 48.26$ c/d detected in the HARPS spectra of HD 170699. The total amount of the phase shifts suggests that it should be a $\ell = 16 \pm 1$ mode [7]. This frequency is the second detected in the spectroscopic data and the 374th in the CoRoT photometric light curve, with an amplitude of about 50 μmag. Therefore, the case of HD 170699 seems to support the conclusion made in the case of HD 50844 [5], i.e., that cancellation effects are not sufficient in completely removing the variations of the integrated flux associated with $\ell \geq 5$ modes. Two other physical mechanisms can be invoked to explain the rich frequency spectra observed in δ Sct stars: granulation effects [1] or island modes, e.g., [2]. These three different scenarios contribute to make the modeling of δ Sct stars a challenging task for theoreticians. In particular, the visibility of island modes may be particularly relevant in the case of HD 170699. We measured $v_{\mathrm{eq}} \sin i = 270\ \mathrm{km\,s^{-1}}$, a value very close to the break-up velocity, and we can infer that HD 170699 is seen equator-on.

References

1. Kallinger, T., Matthews, J.M.: ApJ **711**, L35 (2010)
2. Lignières, F., Georgeot, B.: A&A **500**, 1173 (2009)
3. Mantegazza, L.: In: Breger, M., Montgomery, M.H. (eds.) Delta Scuti Stars and Related Stars, ASP Conf. Series vol. **210**, p. 38 (2000)
4. Mantegazza, L., Poretti, E., Zerbi, F.M.: A&A **366**, 547 (2001)
5. Poretti, E., Michel, E., Garrido, R., et al.: A&A **506**, 85 (2009)
6. Raskin, G., van Winckel, H., Hensberge, H., et al.: A&A **526**, A69 (2011)
7. Telting, J.H., Schrijvers, C.: A&A **317**, 723 (1997)
8. Zima, W.: Comm. Asteroseismol. **155**, 17 (2008)

Recent Progress and Open Questions in Rapidly Rotating Star Seismology

F. Lignières

Abstract Interpreting the frequency spectrum of rapidly rotating stars is a long standing unsolved problem of stellar seismology. The quality of the spatial data together with the advances in modelling the effects of rotation bring fresh perspectives towards a successful comparison between theory and observation. In this short review, the recent progress obtained with 2D oscillation codes is summarized and some open issues are pointed out.

1 Introduction

As of today, the theoretical understanding of stellar oscillations has not been sufficient to perform the detailed seismology of rapidly rotating stars. Since stellar rotation breaks spherical symmetry, the computation of the modes and the physical interpretation of their properties are significantly more complicated in rotating stars as compared to non-rotating stars. Indeed, as eigenfunctions are no longer separable in r and θ, a 2D boundary-value problem in the meridional plane must be solved and interpreted. Historically, rotational effects have been investigated through two approximate methods : the perturbative method, where the rotation rate is assumed to be small $\Omega \ll \omega$ and $\Omega \ll \sqrt{GM/R^3}$, and the traditional approximation for gravito-inertial modes where the centrifugal force and the horizontal component of the rotation vector in the expression of Coriolis force are neglected in the limit $\omega \ll N$ and $\Omega \ll N$. In both cases, the computation and the analysis are much simplified since separability reduces the initial 2D boundary-value problem to 1D boundary-value problems. The first and second order perturbative methods

F. Lignières (✉)
CNRS, IRAP, 14, avenue Edouard Belin, F-31400 Toulouse, France
Université de Toulouse, UPS-OMP, IRAP, 14, avenue Edouard Belin, F-31400 Toulouse, France
e-mail: francois.lignieres@irap.omp.eu

Lignières, F.: *Recent Progress and Open Questions in Rapidly Rotating Star Seismology.* Astrophys Space Sci Proc. **31**, 43–48 (2013)
DOI 10.1007/978-3-642-29630-7_8, © Springer-Verlag Berlin Heidelberg 2013

are well adapted for the Sun and for slowly rotating stars like most solar-type p-mode pulsators. However, these approximate methods are questionable for typical upper-main-sequence pulsators as well as for the most rapidly rotating solar-type pulsators.

Starting with Clement [7], there have been several attempts to compute accurate numerical solutions of the full 2D boundary-value problem, without approximation on the effect of the Coriolis force and the centrifugal deformation of the star on the oscillations. This paper is a short review of the recent progress and the open questions that came out from 2D oscillation computations. After a brief description of the construction and the use of 2D oscillation codes (Sect. 1), I summarize results on the validity of the approximate rotation treatments (Sect. 2), the asymptotic properties of pressure modes (Sect. 3), the importance of mode coupling (Sect. 4) and the current efforts to build tools for mode identification (Sect. 5).

2 Constructing and Using 2D Oscillation Codes

Among the existing 2D stellar oscillation codes, some include a full treatment of the Coriolis and the centrifugal forces [9, 22, 25], while others consider either the Coriolis force [11, 31] or the centrifugal force alone [15], or the Coriolis force plus a first order approximation to the centrifugal deformation [14]. Effects of the differential rotation have been included in [20,27]. They also distinguish themselves by the choices of the discretisation method, the algebraic eigenvalue solver, the coordinate system and the dependent variables. The accuracy and efficiency of the code depend on all these choices and, as illustrated by the series of Clement's paper [7–9], controlled and small numerical errors are not guaranteed. Low-degree p and g modes of low and high frequencies in a uniformly rotating polytropic model of stars have been computed with high accuracy using the TOP code [2, 25]. These calculations have been used as a benchmark to test other oscillation codes (see [22]).

The numerical exploration of the mode properties using 2D oscillation codes presents some specific difficulties. First, and contrary to the non-rotating case, the code searches for solutions in a crowded frequency spectrum corresponding to all the modes of the same symmetry class, that is modes of the same azimuthal number m and equatorial parity. This implies that modes of various wavelengths are computed at the same time, so that in practice one has to take care of poorly resolved modes computed together with well-resolved ones. Another difficulty concerns the physical classification of the modes since, contrary to non-rotating case, their spatial distribution can no longer be characterized by the numbers of respectively radial and latitudinal nodes. In this context, a method of numerical exploration that proved very useful is to choose a set of modes at zero rotation and to follow them by progressively increasing the rotation rate [2, 8, 10, 15, 19, 25] (see also F. Peña's contribution). It must nevertheless be kept in mind that a global understanding of the spectrum properties can hardly be drawn from a numerical exploration that is necessarily limited.

3 Testing the Approximate Treatments of the Rotation Effects

The limit of validity of perturbative methods can be determined using 2D oscillation codes. In recent years, this question has been addressed for low-degree ($\ell = 0, 1, 2, 3$) p and g modes in different frequency ranges and for different stellar models, by comparing the 2D results either with a polynomial fit of $\omega(\Omega)$ [2, 15, 25, 26] or with existing perturbative methods [5, 32]. For p modes, the validity range of the second order polynomial fit shortens for higher frequency as $\Omega \propto \Omega_K (\delta\omega/\omega)^{1/2}$ where $\delta\omega$ is the authorized departure from the exact value. This can be understood as due to an error on the dimension of the resonant cavity that is proportional to $(\Omega/\Omega_K)^2$. Comparison with existing perturbative methods showed that the inclusion of second-order near-degeneracy effects is essential to extend their domain of validity. It goes up to $\Omega/\Omega_K \sim 0.09$ for high-frequency p modes of a solar-type pulsators [32]. For g modes the main feature is the abrupt breakdown of the second order polynomial fit in the sub-inertial regime $\omega < 2\Omega$. Again, this is attributed to the modification of the resonant cavity that is not captured by perturbative methods [2].

Recently, Ballot et al. [3] have used their 2D calculation of low-degree g modes in rotating $N = 3$ polytropic models to test the traditional approximation and found relatively good agreement despite the fact that the traditional approximation does not take into account the centrifugal deformation. However, a detailed study is required to further investigate high-frequency g modes and effects of a significant convective zone. The calculations by Saio et al. [30] also pointed out a striking difference on the results of mode linear stability analysis with and without the traditional approximation.

4 The Asymptotic Properties of p Modes: Numerical Computation and Ray-Based Theory

The follow-up of low-degree high-frequency p modes up to high rotation rates showed that the Tassoul-like spectrum organisation [6] is replaced by a new organisation that is also characterized by constant frequency spacings [15, 26, 27]. The spectrum modification is accompanied by drastic changes in the geometry of the mode resonant cavity. These empirical findings have been interpreted in the context a ray-based asymptotic theory that has been constructed using concepts and techniques developed in quantum physics [16, 17]. Accordingly, the high-frequency p-mode spectrum is the superposition of subspectra associated with dynamically independent families of acoustic rays. At zero rotation, there is only one family corresponding to whispering gallery rays, but as rotation increases the ray dynamics undergoes a smooth transition towards a mixed system where chaotic regions of phase space coexist with regular ones. New type of modes are associated with rays that remain in the vicinity of the stable periodic orbits, they are called island modes because the rays appear as islands in the Poincaré surface of section. The low-degree high-frequency p modes that had been followed from zero rotation are indeed associated with the shortest stable periodic orbits. Semi-analytical solutions that relate their potentially observable regular spacings to internal physical properties have been found [23]. Another type of mode whose relative number increases with rotation is associated with chaotic rays. These modes are thus called chaotic modes and they have specific properties like a generic statistical distribution. But it should be clear that only rays are chaotic while the modes made from constructive interferences of these rays do not show deterministic chaos features. This asymptotic picture has been successfully confronted with $\sim$ 300 high-frequency numerically computed p modes.

While the numerical exploration provides an accurate but necessarily limited view of the full spectrum, the ray-based asymptotic theory gives an approximated but global and physical understanding. A ray-based asymptotic theory has been also developed for gravito-inertial modes; the first results are published in [3].

5 Mode Coupling

In non-rotating stars, avoided crossings between g and p modes of the same ℓ and m lead to mixed modes. In rotating stars, such mode coupling can occur between all modes of the same symmetry class (same m and equatorial symmetry) and it is actually a common feature because the spectrum of a given symmetry class is modified by rotation and the frequencies tend to cross each other [15]. As for mixed modes, mode properties like visibility or stability can change drastically during the avoided crossing. A striking example is the stability of low-frequency g modes in B stars analyzed in detail in [1]. It is shown that rotation-induced coupling with strongly damped modes stabilizes retrograde modes while prograde modes are not much affected by coupling and remain unstable. A systematic study of the occurrence and the relevance of avoided crossings for p and g modes has yet to be performed.

Another important open issue is how rotation impacts non-linear mode coupling and, as a consequence, mode amplitudes. The resonant weakly non-linear interaction between an unstable mode and two stable modes has indeed been proposed as a possible non-linear saturation mechanism that determines mode amplitudes in classical pulsators. While in non-rotating stars this resonance is only possible between modes satisfying $\ell_1 + \ell_2 = \ell_3$, in rotating stars a resonant coupling can occur between all modes of the same symmetry class. In addition, the rotational frequency splitting increases the possibility of fulfilling the resonance condition between the mode frequencies. As shown in [12, 21] this last effect leads to the decrease of the unstable mode amplitude as a function of rotation. We might expect that the relaxed condition on spatial coupling has a similar effect. While challenging, a study of the rotation effects on non-linear mode coupling could help to understand the striking decrease of mode amplitude with rotation found in δ Scuti stars [4]. It might also play a role in the high number of modes observed in δ Scuti with CoRoT [24].

6 Towards Mode Identification

Even without amplitude predictions, identification constraints might be obtained from the detection of regular frequency patterns. Indeed, evidences of constant frequency spacings have been found with CoRoT and MOST in δ Scuti [13] and Be stars [6]. A recent attempt to look for regular spacing in p-mode synthetic spectra calculated from the asymptotic theory and approximated visibilities indicates that a quantity similar to the large separation and the 2Ω frequency spacing between $m = 1$ and $m = -1$ modes are the most likely features to be detected. Then, depending on the inclination angle, half the large separation and an equivalent of the small separation although not small could be also detected [18]. Ballot et al. [3] also found regular patterns among low-degree g modes of a $N = 3$ polytropic model, that are generally well described by the traditional approximation. As these patterns are not uniform spacings, a method of analysis must be proposed to detect them in a spectrum. A third type of regular patterns concerns the gravity modes or inertial modes with very low frequency in the co-rotating frame $\omega \ll \Omega$. For the modes excited by the iron opacity bump, this results in the grouping of frequencies of the same azimuthal number m in the observer's frame $\omega_i = \omega - m\Omega$, a property that is observed in Be stars [6].

In order to constrain the mode identification with spectroscopic data, the method of analysis of the line profile variations must be adapted to rapid rotators. This has been done using approximate treatment of rotation effects but not with modes resulting from 2D calculations.

7 Conclusion

2D oscillation codes are crucial for improving our understanding of rotation effects. The seismology of rapidly rotating stars should benefit from this modelling progress and the analysis of CoRoT and Kepler data. Our inability to predict mode amplitudes in upper-main-sequence pulsators remains the main modelling obstacle. The construction of realistic internal structure models that takes into account rotational effect including the centrifugal deformation is another difficulty although efforts are underway [28,29]. From the observational point of view, it must stressed that constraints on the rotation rate of Kepler and CoRoT stars are essential in order to investigate the evolution of the pulsational behaviour with rotation.

Acknowledgements I would like to thank support from the ANR SIROCO and J. Ballot and M. Pasek for fruitful discussions in preparing this review.

References

1. Aprilia, Lee, U., Saio, H.: MNRAS **412**, 2265 (2011)
2. Ballot, J., Lignières, F., Reese, D.R., et al.: A&A **518**, A30 (2010)
3. Ballot, J., Lignières, F., Prat, V., et al.: to appear in ASP proceedings of "The 61st Fujihara seminar: Progress in solar/stellar physics with helio- and asteroseismology" held in Hakone on March 2011 (2011) arXiv:1109.6856
4. Breger, M.: CoAst **150**, 25 (2007)
5. Burke, K.D., Reese, D.R., Thompson, M.J.: MNRAS **414**, 1119 (2011)
6. Cameron, C., Saio, H., Kuschnig, R., et al.: ApJ **685**, 489 (2008)
7. Clement, M.J.: ApJ **249**, 746 (1981)
8. Clement, M.J.: ApJ **301**, 185 (1986)
9. Clement, M.J.: ApJS **116**, 57 (1998)
10. Deupree, R.G., Beslin, W.: ApJ **721**, 1900 (2010)
11. Dintrans, B., Rieutord, M.: A&A **354**, 86 (2000)
12. Dziembowski, W., Krolikowska, M., Kosovichev, A.: AA **38**, 61 (1988)
13. García Hernández, A., Moya, A., Michel, E., et al.: A&A **506**, 79 (2009)
14. Lee, U., Baraffe, I.: A&A **301**, 419 (1995)
15. Lignières, F.: Rieutord, M., Reese, D.: A&A **455**, 607 (2006)
16. Lignières, F., Georgeot, B.: Phys. Rev. E **78**, 016215 (2008)
17. Lignières, F., Georgeot, B.: A&A **500**, 1173 (2009)
18. Lignières, F., Georgeot, B., Ballot, J.: AN **331**, 1053 (2010)
19. Lovekin, C.C., Deupree, R.G.: ApJ **679**, 1499 (2008)
20. Lovekin, C.C., Deupree, R.G., Clement, M.J.: ApJ **693**, 677 (2009)
21. Nowakowski, R.M.: AA **55**, 1 (2005)
22. Ouazzani, R.: PhD Thesis (2011)
23. Pasek, M., Georgeot, B., Lignières, F., et al.: Phys. Rev. Lett. **107**, 121101 (2011)
24. Poretti, E., Michel, E., Garrido, R., et al.: A&A **506**, 85 (2009)
25. Reese, D., Lignières, F., Rieutord, M.: A&A **455**, 621 (2006)
26. Reese, D., Lignières, F., Rieutord, M.: A&A **481**, 449 (2008)
27. Reese, D.R., MacGregor, K.B., Jackson, S., et al.: A&A **506**, 189 (2009)
28. Rieutord, M., Espinosa Lara, F.: CoAst **158**, 99 (2009)
29. Roxburgh, I.W.: A&A **454**, 883 (2006)
30. Saio, H., Cameron, C., Kuschnig, R., et al.: ApJ **654**, 544 (2007)
31. Savonije, G.J.: A&A **469**, 1057 (2007)
32. Suárez, J.C.: Goupil, M.J., Reese, D.R., et al.: ApJ **721**, 537 (2010)
33. Tassoul, M.: ApJS **43**, 469 (1980)

Traditional Approximation for Low-Frequency Modes in Rotating Stars and A Working Hypothesis About Episodic Mass Loss in Be Stars

Hiromoto Shibahashi and Hiroyuki Ishimatsu

Abstract The so-called traditional approximation is reasonably good for very low frequency modes of rotating stars. In this approximation, the angular dependence of eigenfunction is expressed in terms of the Hough functions, and the radial dependence is expressed in a form similar to the case of non-rotating stars with only the replacement of the spherical degree with the eigenvalue of the Hough function. Paying attention to the non-zero surface temperature, we point out that very low frequency oscillations become leaky waves. A working hypothesis is then proposed to explain mass loss in Be stars as episodic wave leakage associated with gradual angular-momentum transport due to prograde modes.

1 Traditional Approximation for Low-Frequency Modes in Rotating Stars

In a rotating star, the Coriolis force modifies significantly the oscillation characteristics of the star, particularly those of low-frequency modes. The "traditional approximation," neglecting the horizontal component of the Coriolis force associated with radial motion and the radial component of the Coriolis force associated with horizontal motion, is helpful to simplify the mathematical description keeping the essence of the physics (see, e.g., [4]). It is reasonably good indeed for very low frequency modes, for which the horizontal motion dominates the oscillation. In this approximation, the azimuthal eigenfunction is expressed by $\exp(im\phi)$, where ϕ denotes the azimuth angle in polar coordinates and m is an integer, and

H. Shibahashi (✉) · H. Ishimatsu
Department of Astronomy, University of Tokyo, Bunkyo-ku, Tokyo 113-0033, Japan
e-mail: shibahashi@astron.s.u-tokyo.ac.jp; ishimatsu@astron.s.u-tokyo.ac.jp

Shibahashi, H. and Ishimatsu, H.: *Traditional Approximation for Low-Frequency Modes in Rotating Stars and A Working Hypothesis About Episodic Mass Loss in Be Stars.* Astrophys Space Sci Proc. **31**, 49–52 (2013)
DOI 10.1007/978-3-642-29630-7_9,

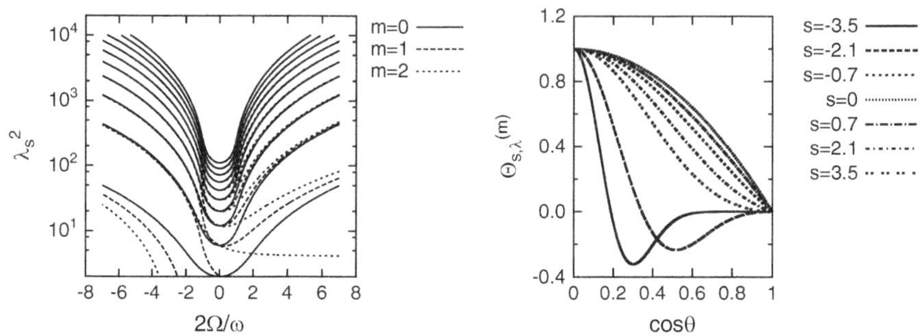

Fig. 1 *Left*: Eigenvalues of Laplace's tidal equation λ_s^2 versus the rotation parameter $s = 2\Omega/\omega$. The azimuthal and temporal dependence is given by $\exp(im\phi - i\omega t)$. *Right*: The Hough functions in the case of $m = 2$ for some values of s versus cosine of colatitude. Only those functions are plotted which tend to a quadrupole ($l = 2$) mode when Ω tends to zero

the latitudinal dependence is expressed in terms of the Hough functions, which are eigensolutions of the Laplace tidal equation [3, 5]. The eigenvalues, λ_s^2, of the Laplace tidal equation, which is described by the azimuthal order m and the rotation parameter $s \equiv 2\Omega/\omega$, Ω being the rotation frequency and ω the oscillation frequency, tend to $l(l+1)$ in the limit of the non-rotating case, where l denotes the spherical degree, and they deviate substantially from $l(l+1)$ with increasing rotation parameter, as shown in the left panel of Fig. 1. The Hough functions naturally tend to the associated Legendre functions in the limit of the non-rotating case. With increasing rotation frequency, the Hough functions, particularly those of retrograde modes, tend to have relatively much larger amplitude in the low-latitude zone than in the high-latitude region, as seen in the right panel of Fig. 1.

The radial part of the eigenfunctions is governed by equations having the same form as in the case of a non-rotating star in the Cowling approximation, in which the Eulerian perturbation to the gravitational potential is neglected, but with $l(l+1)$ replaced by λ_s^2. We can then discuss some qualitative properties of the low-frequency oscillations of a rotating star immediately in a quite similar way to discussions on a non-rotating star, only by replacing $l(l+1)$ with λ_s^2. The global nonradial oscillations are standing waves which are trapped in a propagative region by the reflection of the waves at both sides of the region. The qualitative features of wave trapping can be understood with a local dispersion relation, which relates the wave number in the radial direction, k_r, to the frequency, ω [9];

$$k_r^2 = \omega^{-2}c^{-2}(\omega^2 - N^2)(\omega^2 - \lambda_s^2 c^2/r^2), \tag{1}$$

where c denotes the sound speed and N the Brunt-Väisälä frequency. Waves can propagate in the radial directions only in regions where $\omega^2 < N^2, \lambda_s^2 c^2/r^2$ or $\omega^2 > N^2, \lambda_s^2 c^2/r^2$, and in other regions the waves are evanescent.

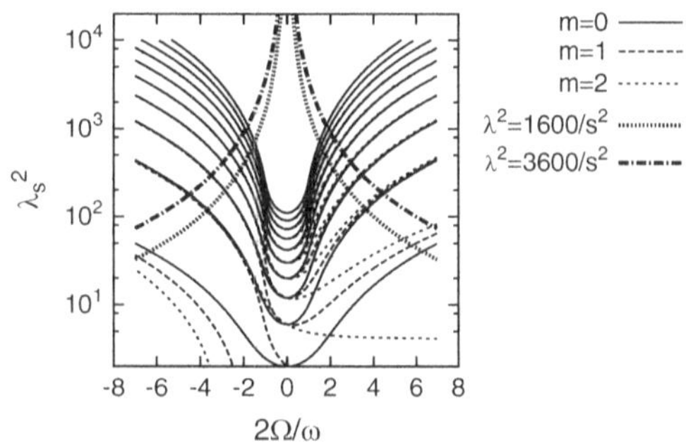

Fig. 2 The critical eigenvalues of λ_s^2 in the case of $m = 2$ for $v_{eq} = 200\,\mathrm{km\,s^{-1}}$ and $v_{eq} = 300\,\mathrm{km\,s^{-1}}$ versus the rotation parameter $2\Omega/\omega$. The azimuthal and temporal dependence is given by $\exp(im\phi - i\omega t)$

2 Surface Critical Cut-Off Frequency

In the case of a star having a thick convective envelope, such as a $1\,M_\odot$ main-sequence star, low-frequency waves are evanescent in the envelope and hence the surface amplitude must be tiny even if the waves would be self-excited. Only in the case of stars having a radiative envelope, the surface amplitude of low-frequency modes is expected to be large. However, if $\omega^2 < \lambda_s^2 c^2/r^2$ at the surface, that is, if $|\omega| < \lambda_s c(R)/R$, no reflection is expected at the surface and the waves leak out. Hence, in order to for the wave to be excited as a standing wave,

$$|s|\lambda_s = (2\Omega/|\omega|)\lambda_s < 2v_{eq}/c(R) \tag{2}$$

is necessary, where v_{eq} denotes the equatorial rotational velocity of the star. Figure 2 shows the critical values of λ_s^2 as functions of the rotation parameter $s = 2\Omega/\omega$ for the cases of $v_{eq} = 200\,\mathrm{km\,s^{-1}}$ and $v_{eq} = 300\,\mathrm{km\,s^{-1}}$ with $c(R) = 10\,\mathrm{km\,s^{-1}}$. Those modes corresponding to values of λ_s above the two thick dashed lines in Fig. 2 are no longer standing waves.

3 A Working Hypothesis for Episodic Mass Loss in Be Stars

Be stars eject their mass from the equatorial region quasi-periodically at intervals ranging from several years to some decades, and this mass forms a cool gas disc around them. The mechanism of the episodic mass loss in Be stars is, however, as yet unknown [2, 8]. Here we discuss the relation between nonradial oscillations and mass loss in Be stars from a point of view of oscillations in a rotating star. Actually, many Be stars are nonradial oscillators with long periods, and some Be stars are rotating very close to critical velocity.

It should be noted first that the non-axisymmetric nonradial oscillations can transport angular momentum from the driving zone to the damping zone. If low-frequency prograde nonradial oscillations are selectively excited by the

κ-mechanism of the iron bump, which seems to work in β Cep and SPB stars, angular momentum transported by nonradial oscillations is deposited near the surface due to dissipation. This will result in a gradual increase in λ_s, leading eventually to break the standing wave condition (2). Once nonradial oscillations start to leak out, the damping effect becomes large so that a big phase shift appears between the radial and azimuthal components of the oscillation velocity. The angular momentum flux across the surface is roughly proportional to sine of this phase shift. This angular momentum transport is closely related to the mechanical energy flux of waves and, in the case of leaky waves, the angular momentum is transported by prograde waves outward [6, 7]. Therefore it is expected that the angular momentum is transported outward with increasing speed once the waves become leaky. On the other hand, the oscillation itself will be soon damped owing to kinetic energy loss overcoming the excitation mechanism. Also, once angular momentum is lost from the star, the envelope of the star spins down and then the condition (2) is recovered. Hence angular momentum loss stops very soon. The star remains quiet until new nonradial oscillations are built up by the κ-mechanism to sufficient amplitude and a new episode begins. According to this scenario, the interval of episodic mass loss corresponds to the growth time of the oscillation. It is expected to be of the order of a decade or longer, and seems to be fairly in good agreement with observations.

This model requires prograde nonradial oscillations is predominant, at least in a certain phase, in Be stars. Though excitation has not yet explicitly been shown, it has been found that rapid rotation tends to stabilize retrograde g-modes while prograde sectoral modes are exceptionally immune to the damping effects [1].

Acknowledgements This research was partially supported by the Japan Society for the Promotion of Science, Grant-in-Aid for Scientific Research (C) 23540260.

References

1. Aprilia, Lee, U., Saio, H.: Stability of g modes in rotating B-type stars. MNRAS **412**, 2265–2276 (2011)
2. Balona, L.A.: Rotational modulation in Be stars. In: Suárez, J.C., Garrido, R., Balona, L.A., Christensen, J. (eds.) Stellar Pulsations. Springer, Heidelberg (2012)
3. Bildsten, L., Ushomirsky, G., Cutler, C.: Ocean g-modes on rotating neutron stars. ApJ **460**, 827–831 (1996)
4. Eckart, C.: Hydrodynamics of Oceans and Atmospheres. Pergamon Press, Oxford (1960)
5. Lee, U., Saio, H.: Low-frequency nonradial oscillations in rotating stars. I. Angular dependence. ApJ **491**, 839–845 (1997)
6. Osaki, Y.: Nonradial pulsation theory of massive stars. PASP **98**, 30–32 (1986)
7. Pantillon, F.P., Talon, S., Charbonnel, C.: Angular momentum transport by internal gravity waves III. Wave excitation by core convection and the Coriolis effect. A&A **474**, 155–163 (2007)
8. Rivinius, T.: Be stars: Rapidly rotating pulsators. In: Suárez, J.C., Garrido, R., Balona, L.A., Christensen, J. (eds.) Stellar Pulsations. Springer, Heidelberg (2012)
9. Unno, W., Osaki, Y., Ando, H., Saio, H., Shibahashi, H.: Nonradial Oscillations of Stars. University of Tokyo Press, Tokyo (1989)

Modelling Stellar Convection and Pulsation in Multidimensions Using the ANTARES Code

Eva Mundprecht and Herbert J. Muthsam

Abstract The ANTARES code has been designed for simulation of astrophysical flows in a variety of situations, in particular in the context of stellar physics. Here, we describe extensions as necessary to model the interaction of pulsation and convection in classical pulsating stars. These extensions comprise the introduction of a spherical grid, moveable in the radial direction, specific forms of grid-refinement and considerations regarding radiative transfer. We then present the basic parameters of the Cepheid we study more closely. For that star we provide a short discussion of patterns of the H + He I and the He II convection zones and the interaction with pulsation seen in the pdV work or atmospheric structures.

1 Introduction

Time-dependent, non-linear models of the classical pulsating variables (Cepheids, RR Lyr) have been around for decades. Nearly all those models have treated the problem in one spatial dimension (plus time). As is well known, for the stars near the red edge the inclusion of convection is mandatory. Given the basic 1D computational setting, this had to be done using simplified models for convection. For a detailed discussion of shortcomings of this approach and their consequences see [1, 2].

We therefore have adapted the ANTARES code, [5], in order to meet the needs for modelling such stars in multidimensions (2D presently). In Sect. 2 we address a few technical issues and in Sect. 3 we discuss a specific model.

E. Mundprecht (✉) · H.J. Muthsam
Faculty of Mathematics, University of Vienna, Austria
e-mail: eva.mundprecht@univie.ac.at; herbert.muthsam@univie.ac.at

Mundprecht, E. and Muthsam, H.J.: *Modelling Stellar Convection and Pulsation in Multidimensions Using the ANTARES Code*.
Astrophys Space Sci Proc. **31**, 53–56 (2013)
DOI 10.1007/978-3-642-29630-7_10,

2 Technical Issues

The simulations were performed on a stretched, polar and moving grid, covering a sector in latitude. We discuss the radiative transfer equation (RTE) and restrictions on time stepping in turn.

The Radiative Transfer Equation. Near the surface, nontrivial energy exchange between gas and radiation is included via the RTE with grey opacities, while the diffusion approximation is used for the deeper layers. The RTE in 1D is solved along single rays via the short characteristics method of Kunasz and Auer [4]. In 2D either 12 or 24 ray directions are chosen according to the angular quadrature formulae of type A4 or A6 of Carlson [3], and the directions in each quadrant are arranged in a triangular pattern. For each ray the points of entrance and exit plus the corresponding distances are determined. Since the grid moves this has to be redone every step. The RTE

$$\mu \frac{\partial I}{\partial \tau} = I - S \tag{1}$$

is then solved along each ray. Here I denotes the intensity, S the source function, and τ the vertical optical depth scale, finally $\mu = \cos\theta$, where θ is the polar angle. This procedure is repeated recursively since after the first step one gets just the intensity on a single new point.

Time Stepping. The time step for our (explicit) method is never limited by the classical CFL condition but is due to radiation transport. Using the diffusion approximation also near the surface would enforce prohibitively small time steps,

$$\Delta t_{\text{diff}} \propto \min\left(3c_p/\left(16\kappa\sigma T^3\right)[\kappa\rho/k]^2\right) \tag{2}$$

Fortunately, the radiative transfer equation allows the time step to stay in contact with the time scale for relaxing a temperature perturbation on the scale of the grid size by radiation (Spiegel's relation, [6]),

$$\Delta t_{\text{rad}} \propto \min\left(c_p/\left(16\kappa\sigma T^3\right)\left[1-(\kappa\rho/k)\,\text{arccot}\,(\kappa\rho/k)\right]^{-1}\right), \tag{3}$$

with $k = 2\pi/\min(\Delta r_i, \Delta y_i)$, which in practice leads to larger time steps.

3 Results

3.1 *Model Settings*

The parameters of our Cepheid are $T_{\text{eff}} = 5125$ K, $\log g = 1.97$, $L = 913\ \text{L}_\odot$ and $M = 5\ \text{M}_\odot$. The period is about 4 days, the radius is 26 Gm, of which the outer 11.3 Gm were modelled. The opening angle of the sector calculated is either

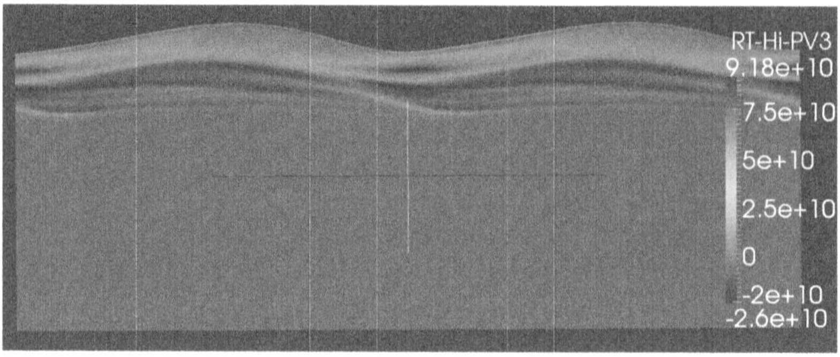

Fig. 1 *pdV* work due to convection, averaged over ten periods as a function of pulsational phase. Below the convection zone proper, overshooting plumes also contribute to some extent

1° or 10°. For the 10° model the computational area has 510 radial grid points and 800 lateral points. For the 1° model these number are 800 and 300, respectively. Radially the grid is stretched from cell to cell by a factor of 1.011 and 1.07, respectively. The radial mesh sizes of the 10° model vary from 0.47 Mm at the top to 124 Mm at the bottom. The runs with these opening angles were started from one and the same 1D model which had been relaxed for more than 100 periods. The 10° model is wide enough to harbour convection cells of the He II convection zone. Its resolution is insufficient to reasonably represent the H + He I convection zones. This purpose is achieved by the 1° model. We discuss now the He II and the H + He I convection zone in turn.

3.2 *The He II Convection Zone*

After some time granted to the He II convection zone for proper development, ten useful periods for evaluation are at our disposal. Visualization shows a clear dependence of convection on phase. Near maximum compression the lower part of descending plumes tends to detach from the body of the c.z., and upon expansion the body reorganizes, gains strength and new plumes appear. Figure 1 shows phase dependence for the *pdV* work of the He II zone, averaged over the ten useful periods. *pdV* refers to the work done by convection *only*, without pulsational contribution. Besides the varying strength of the work we see also some contributions due to plumes plunging into the stable region below. These plumes also generate noticeable gravity waves, discernible down to our lower boundary. Laterally, their extent coincides with the width of our computational domain. An even wider sector may be appropriate. The plumes have not yet reached their statistically steady state but keep growing. With such models, it is also possible to test and calibrate, for example, convection models routinely used in modelling radially pulsating stars.

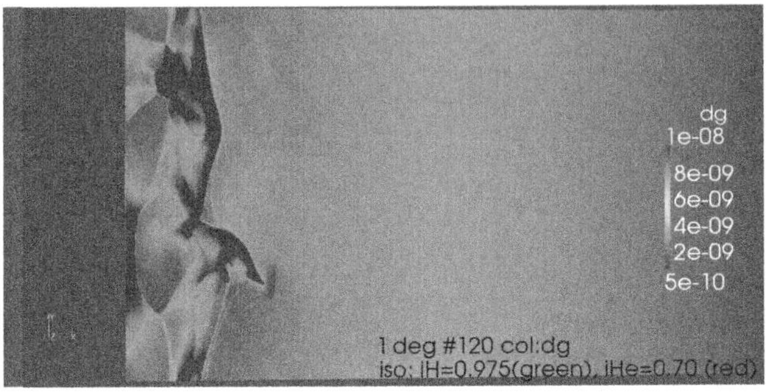

Fig. 2 Upper convection zone upon contraction. Top is to the left. Colours: density of gas. Isolines mark the indicated ionization degrees of hydrogen and helium

3.3 The H + He I Convection Zone

With the 1° model illustrated in Fig. 2 the H + He I is reasonably resolved and, as a result, much more vigorous than in the 10° case. Velocities (pulsation subtracted) easily get supersonic. Note the network of shocks above the hydrogen ionization front.

Frequently, shocks move upwards. In one case, this leads to the strong high-density region quite at the top. Here, our closed boundary conditions failed. We are working on implementing open boundary conditions at the top. Mass loss in Cepheids is a topic of present discussion. Results such as these hint at the intriguing possibility that convectively induced shocks may be a cause. Generally, it should be noted that such calculations are very expensive even in 2D and by no means all Cepheids are within computational reach.

Acknowledgements We acknowledge support by the Austrian Science Foundation (P20973).

References

1. Buchler, J.R.: In: Ferlet, R., Maillard, J.P., Raban, B. (eds.) Variables Stars and the Astrophysical Returns of the Microlensing Surveys, p. 181 (1997)
2. Buchler, J.R.: In: American Institute of Physics Conference Series, vol. 1170, 51 (2009)
3. Carlson, B.G.: In: Alder, B., Fernbach, S. (eds.) Methods in Computational Physics, vol. 1 (1963)
4. Kunasz, P., Auer, L.H.: JQSRT **39**, 67 (1988)
5. Muthsam, H.J., Kupka, F., Löw-Baselli, B., et al: NewAst **15**, 460 (2010)
6. Spiegel, E.A.: ApJ **126**, 202 (1957)

Regular Modes in Rapidly Rotating Stars

Mickael Pasek, Bertrand Georgeot, François Lignières, and Daniel R. Reese

Abstract Rapid rotation has to date prevented reliable mode identification in most intermediate to massive pulsating stars. In this contribution we show how an asymptotic method for pressure modes predicts regular frequency spacings in the spectrum of rapidly rotating stars. We give a semi-analytical formula for these spacings that is in good agreement with numerically computed high-frequency modes. This should help to extract from observed spectra important information on the structure of rapidly rotating stars.

1 Introduction

In the field of asteroseismology the rapid rotation of a star is still a theoretical obstacle to successful mode identification of pulsation frequencies. It has recently been found empirically in numerical models [4, 7] that a subset of high-frequency pressure modes of rapidly rotating stars show regular frequency spacings, whether computed for polytropic or more realistic models [8]. On the observational side

M. Pasek (✉)
Université de Toulouse, CNRS, IRAP, 14 avenue Edouard Belin, F-31400 Toulouse, France

Université de Toulouse, CNRS, Laboratoire de Physique Théorique (IRSAMC), F-31062 Toulouse, France
e-mail: pasek@irsamc.ups-tlse.fr

B. Georgeot
Université de Toulouse, CNRS, Laboratoire de Physique Théorique (IRSAMC), F-31062 Toulouse, France

F. Lignières
Université de Toulouse, CNRS, IRAP, 14 avenue Edouard Belin, F-31400 Toulouse, France

D.R. Reese
LESIA, CNRS, Université Pierre et Marie Curie, Université Denis Diderot, Observatoire de Paris, 92195 Meudon Cedex, France

Pasek, M. et al.: *Regular Modes in Rapidly Rotating Stars.*
Astrophys Space Sci Proc. **31**, 57–60 (2013)
DOI 10.1007/978-3-642-29630-7_11,

some hints of regularities have been found in the spectra of rapidly rotating δ Scuti stars [2]. Knowing that in non-rotating stars the p-mode frequency spacings can be described by the Tassoul asymptotic formula [9], a natural question to ask is whether one can derive an asymptotic formula for the regular spacings that are expected to be seen in the p-mode spectrum of rapidly rotating stars. We have very recently shown [6] that it is indeed possible and explain how in the following.

The Tassoul asymptotic formula at leading order reads

$$\omega_{n,\ell} \simeq \Delta\omega \left(n + \frac{\ell}{2} + \frac{1}{4} + \alpha \right), \tag{1}$$

with the large separation

$$\Delta\omega = 2\pi \left(2 \int_0^R \frac{dr}{c(r)} \right)^{-1}, \tag{2}$$

where $c(r)$ is the sound speed that depends on the radial coordinate only. This formula can be derived from the WKB approximation on the radial part of the mode using explicitly the spherical symmetry of a non-rotating star [9]. On the other hand the short-wavelength approximation applied on the three or two-dimensional wave equation leads to a propagation equation for acoustic rays [3]. In a rapidly rotating star the spherical symmetry is no longer valid so the problem cannot be reduced to one dimension only. To circumvent this problem we studied the dynamics of the ray limit. This ray system is of a mixed type which means that rays can be regular or chaotic depending on their initial conditions [5]. It is known from semiclassical theory and quantum chaos that to the regular structures of phase space should be associated a whole family of modes with a regular spectrum.

2 From Stable Rays to Gaussian Beams and Pressure Modes

The method we use was first developed by Babich and is described in [1]. By deriving an approximation of the wave equation in the close vicinity of a stable periodic ray and finding its solution, one can obtain approximate eigen-modes and frequencies for the subset of modes associated with the phase space region of the stable periodic ray.

We start from the equation describing small adiabatic perturbations of a polytropic equilibrium model using the Cowling approximation. Since we are interested in the high frequency range, the Coriolis force and the term corresponding to gravity waves can be discarded. From the axial symmetry of the star one can write the mode amplitude Ψ as $\Psi = \Psi_m \exp(im\phi)$, with ϕ the azimuthal coordinate, to obtain a Helmholtz equation in the meridian plane for $\Phi_m = \sqrt{d}\,\Psi_m$ as

$$\Delta\Phi_m + k^2(\mathbf{r})\Phi_m = 0, \tag{3}$$

with

$$k^2(\mathbf{r}) = \frac{\omega^2}{c_s^2}\left[1 - \frac{\omega_c^2}{\omega^2} - \frac{c_s^2(m^2 - \frac{1}{4})}{\omega^2 d^2}\right] = \frac{\omega^2}{\tilde{c}_s^2} \tag{4}$$

where ω_c is the acoustic cut-off frequency and d the distance to the rotation axis [5]. To apply the approximation method we start from a stable periodic solution γ of the ray system. Then Eq. (3) must be written in the local frame of this stable ray using the longitudinal coordinate s along the ray and the transverse coordinate n perpendicular to it. One then assumes that the mode scales as $s = O(1)$, $n = O(1/\sqrt{\omega})$. An expansion in inverse powers of ω, with $\nu = \sqrt{\omega}\, n$, yields

$$\frac{\partial^2 V_m}{\partial \nu^2} - K(s)\nu^2 V_m + \frac{2i}{\tilde{c}_s}\frac{\partial V_m}{\partial s} = 0, \tag{5}$$

with $K(s) = \tilde{c}_s(s)^{-3}\partial^2\tilde{c}_s/\partial n^2|_{n=0}$ and $V_m = \Phi_m \exp(-i\omega \int_\gamma \tilde{c}_s^{-1} ds)/\sqrt{\tilde{c}_s}$. This equation is the approximated wave equation in the vicinity of γ and one can recognize in it the equation of a quantum harmonic oscillator transverse to γ (first two terms in ν) and its transport along γ. Thus the solutions of this equation at a fixed s are Hermite–Gauss polynomials

$$V_m^\ell(s, \nu) = \left(\frac{\bar{z}}{z}\right)^{\ell/2} H_\ell(\sqrt{\mathrm{Im}(\Gamma)}\nu)\frac{\exp\left[i\frac{\Gamma}{2}\nu^2\right]}{\sqrt{z}} \tag{6}$$

with H_ℓ the Hermite polynomials of order ℓ. The complex quantity $\Gamma(s)$ describes the transverse width of the solution and is related to the deviation of rays close to γ. The variation of this deviation over one period L_γ of γ can be found by computing the monodromy matrix M such that

$$\begin{bmatrix} z(s + L_\gamma) \\ p(s + L_\gamma) \end{bmatrix} = M \begin{bmatrix} z(s) \\ p(s) \end{bmatrix} \tag{7}$$

with $\Gamma(s) = p(s)/z(s)$ and $p(s) = \tilde{c}_s^{-1} dz/ds$. The solution of wave equation (3) must be uni-valued; this leads to a quantization condition that yields a formula for the regular sub-spectrum in rapidly rotating stars such that

$$\omega_{n,\ell,m} = \delta_n(m)n + \delta_\ell(m)\ell + \beta(m), \tag{8}$$

with

$$\delta_n = \frac{2\pi}{\oint_\gamma \tilde{c}_s^{-1} ds}, \quad \delta_\ell = \frac{2\pi N_r + \alpha}{\oint_\gamma \tilde{c}_s^{-1} ds} \quad \text{and} \quad \beta(m) = \frac{\delta_n + \delta_\ell}{2}, \tag{9}$$

with $\oint_\gamma \tilde{c}_s^{-1} ds$ the acoustic time along γ, and N_r and α are related to the stability of γ.

We have compared the results of the semi-analytical formula evaluated from the propagation of acoustic rays with numerical results of [7] for high-frequency p-modes computed in a polytropic stellar model with index $N = 3$. The individual modes were followed from $\Omega/\Omega_K = 0$ to $\Omega/\Omega_K = 0.896$ for $m = \{-1, 0, 1\}$. The modes are labeled by n and ℓ which are the numbers of nodes in the meridian plane. We found in [6] that the semi-analytical regularities and the full computations of high-frequency p-modes are in good agreement.

The interest of formula (8) is to relate potential seismic observables to internal properties of the star. Indeed δ_n depends on the acoustic time along the ray and δ_ℓ depends on the acoustic time and the second transverse derivatives of the speed of sound. This method is also numerically efficient and allows a rapid computation of asymptotic regularities for a given stellar model in order to guide the search for patterns in numerical or observed spectra.

3 Conclusion

In this contribution we have sketched the derivation of an asymptotic method for p-modes that yields a semi-analytical formula for regular spacings in the spectrum of rapidly rotating stars. We obtained a good agreement with numerically computed high frequency modes and think this formula should prove useful for the mode identification in such stars. A detailed account of the method will be available in a subsequent publication.

Acknowledgements We thank CALMIP for the use of their supercomputers. D.R.R. acknowledges support from the CNES.

References

1. Babich, V.M., Buldyrev, V.S.: Short-Wavelength Diffraction Theory. Springer, New York (1991)
2. García Hernández, A., Moya, A., Michel, E., et al.: A&A **506**(1), 79–83 (2009)
3. Gough, D.O.: Linear Adiabatic Stellar Pulsation. In: Zahn, J.-P., Zinn-Justin, J. (eds.) Les Houches Lectures Session XLVIII, p.399. North-Holland, Amsterdam (1993)
4. Lignières, F., Rieutord, M., Reese, D.R.: A&A **455**, 607 (2006)
5. Lignières, F., Georgeot, B.: A&A **500**, 1173–1192 (2009)
6. Pasek, M., Georgeot, B., Lignières, F., Reese, D.R.: Phys. Rev. Lett. **107**, 121101 (2011)
7. Reese, D.R., Lignières, F., Rieutord, M.: A&A **481**, 449 (2008)
8. Reese, D.R., MacGregor, K.B., Jackson, S., et al.: A&A **506**(1), 189–201 (2009)
9. Tassoul, M.: ApJS **43**, 469–490 (1980)

Periodicities Within the Frequency Spectra of δ Scuti Stars Observed by *Kepler*

A. García Hernández, J. Pascual-Granado, A. Grigahcène, K. Uytterhoeven, A. Moya, R. Garrido, J.C. Suárez, J. Gutiérrez-Soto, and S. Martín-Ruiz

Abstract We showed in previous works that a periodic pattern could be found in the frequency set of δ Sct stars observed by *CoRoT*. We related that pattern with a value of the large separation. However, are these stars particular cases?

We present here preliminary results of our work aimed at answering this and related questions. We looked for periodicities in the frequency spectrum of a sub-sample of 153 δ Sct pulsators observed by *Kepler*. We have determined that ~50% of the sub-sample shows a clear pattern. Moreover, we found three objects with almost the same KIC physical parameters showing the same value of the pattern. We suggest that the pattern is a sign of a large-separation-like structure.

1 Previous Results

Periodicities present within the frequency spectra of some pulsating stars have become a useful observable to derive information about the stellar interior. Periodicities appear in the asymptotic regimes of the spectra, where the radial order,

A.G. Hernández (✉) · J. Pascual-Granado · R. Garrido · J.C. Suárez · J. Gutiérrez-Soto · S. Martín-Ruiz
Instituto de Astrofísica de Andalucía, CSIC, Glorieta de la Astronomía, s/n, 18008, Granada, Spain
e-mail: agh@iaa.es

A. Grigahcène
Centro de Astrofísica, Faculdade de Ciências, Universidade do Porto, Rua das Estrelas, 4150-762 Porto, Portugal

K. Uytterhoeven
Instituto de Astrofísica de Canarias (IAC), Tenerife, Spain
Department of Astrofísica, Universidad de La Laguna (ULL), Tenerife, Spain

A. Moya
Departamento de Astrofísica, Centro de Astrobiología (INTA-CSIC), PO BOX 78, 28691 Villanueva de la Cañada, Madrid, Spain

Hernández, A.G. et al.: *Periodicities Within the Frequency Spectra of δ Scuti Stars Observed by* Kepler. Astrophys Space Sci Proc. **31**, 61–65 (2013)
DOI 10.1007/978-3-642-29630-7_12,

n, is much larger than the spherical degree, ℓ [6]. The best known example are the five-minute oscillations of the Sun [7], where a well defined spacing, known as large separation, can be measured: $\Delta\nu_\ell = \nu_{n+1,\ell} - \nu_{n,\ell}$, where $\nu_{n,\ell}$ is the value of the frequency mode determined by the quantum numbers n and ℓ.

δ Sct stars show radial orders of the order of the spherical degrees, so it is not expected to find a large separation pattern within their frequency spectra; regular frequency spacings might, in contrast, arise from rotational splitting. Nevertheless, some efforts have been carried out to look for any kind of spacing [1,4]. With the satellite observations, the presence of periodicities within the frequency spectrum of δ Sct stars has been confirmed. For the frequency set of the *CoRoT* objects HD 174936 [2] and HD 174966 [3], we calculated a Fourier transform (FT), where we could identify the clear sign of a pattern. In [2], we suggested that this pattern would be related to a large-separation structure: a restricted range in the frequency separation between adjacent modes of the same degree, which typically grows towards the asymptotic large separation $\Delta\nu_\ell$ with increasing frequency. The value of this large separation would not be the same as that one measured in the asymptotic regime.

The question now is if these two stars are peculiar cases within the class of δ Sct stars and if the method could be applied to a widely number of objects. To answer these and other questions we are carrying out a study of a large sample of δ Sct stars observed by *Kepler*. We present here the (very) preliminary results of our work.

2 Periodicities in the *Kepler* Sample

We have selected a sub-sample of 153 δ Sct stars from the whole same sample of 750 δ Sct, γ Dor and hybrid pulsators, observed by the *Kepler* satellite and compiled by Uytterhoeven et al. [5]. We followed the procedure described in [2], selecting different subsets containing the highest frequencies in amplitude, levelling out its amplitudes to one and using a FT to identify patterns. Due to the formalism of our Fourier analysis, a pattern should be identified when some of the highest peaks rise up with values corresponding to the spacing and its sub-multiples. An example of how a pattern is identified is shown in Fig. 1.

Our first result is that we can clearly identify a pattern for 50% (77 stars) of the sub-sample. The range for the value of the spacings found varies from around 20 μHz to almost 90 μHz. We observed that a higher number of frequencies obtained for a star does not ensure to find a pattern. From 36 stars with a relatively low number of frequencies (≤ 35), we have detected a pattern for 13 of them. Even more, one object, with only 20 frequencies, shows a clear spacing of $\sim$38 μHz (see Fig. 2).

More interesting are the non-standard cases. There are some power spectra where not only one spacing could be detected but two or even three (KIC 08881697). Other stars present a lot of peaks in its FT, so that several combinations of sub-multiples are possible. A couple of cases where we do not find any periodic pattern are difficult

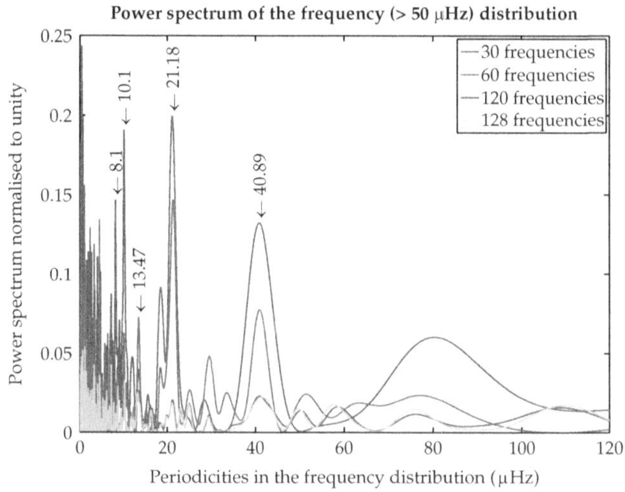

Fig. 1 Fourier transform of different sub-sets of the highest frequencies for the *Kepler* δ Sct star KIC 09551281. The highest peaks are labeled. A peak at 41 μHz, corresponding to the spacing of the pattern, can be seen and also its sub-multiples: 21, 13.5, 10 and 8 μHz. When the number of frequencies increases, the amplitudes of the peaks decrease but never completely disappear

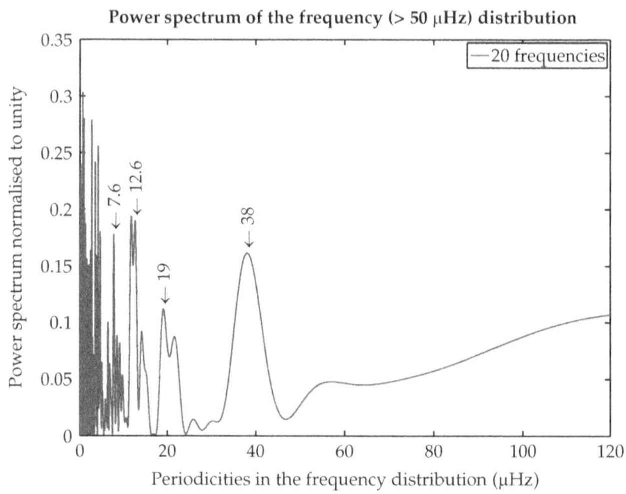

Fig. 2 Fourier transform of the 20 frequencies of KIC 08933391. The spacing of the pattern is 38 μHz, as can be derived from a peak with this value and its sub-multiples

to interpret. KIC 09776474 (see Fig. 3) shows a modulation in its power spectrum which makes the identification of any pattern impossible. This is a strange case, because its FT has a different form from all the other power spectra that we have calculated. All these cases are being investigated in more detail. Finally, we found

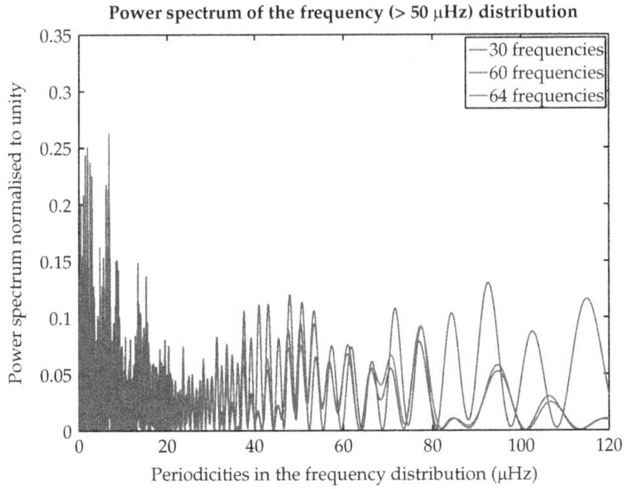

Fig. 3 Power spectrum of the frequency set of KIC 09776474. The modulation makes any identification of a periodic pattern impossible

the same pattern for three analog stars,[1] i.e. with similar $T_{\rm eff}$ and equal $\log g$. This fact points out the relation between the internal structure of the star and the value of the large-separation-like pattern. We suggest that this technique could provide a new observable (together with $T_{\rm eff}$, $\log g$, etc) for δ Sct pulsators.

3 Conclusions

In this work, we are trying to answer the question if the cases of HD 174936 and HD 174966 are peculiar within the group of δ Sct stars. Preliminary results show that there are many of this type of pulsators showing periodic patterns. The *fauna* of the power spectra is quite diverse and some particular cases should be investigated carefully in order to understand them. Some of the patterns found in this preliminary analysis are possibly due to rotational splitting or are spurious, i. e. not coming from the physics of the star. A deeper analysis is necessary to discard false positives and distinguish between these possibilities. Any kind of supporting data (ground-based photometry or spectroscopy) could shed a light on this problem.

Nevertheless, we found the same value of the pattern for three different stars with similar physical parameters. If this pattern is not real, what is the probability of these three stars to have the same random value of the peak in the FT? And, if it is a real pattern, what would be the probability that these stars have the same rotational

[1]We adopted the values of $T_{\rm eff}$ and $\log g$ from the *Kepler* Input Catalog.

velocity (and, hence, the same rotational splitting)? We suggest that the pattern is real in almost all the cases and that it is a sign of a large-separation-like structure. Work is in progress to understand the nature of this pattern and an automatization of the procedure is intended to study large samples of stars.

Acknowledgements Funding for the Kepler mission is provided by NASA's Science Mission Directorate. We thank the Kepler team for the development and operations of this outstanding mission. K.U. acknowledges financial support by the Spanish National Plan of R&D for 2010, project AYA2010-17803. A.M. acknowledges the funding of AstroMadrid (CAM S2009/ESP-1496); this research was funded by the Spanish grants ESP2007-65475-C02-02, AYA 2010-21161-C02-02.

References

1. Breger, M., Pamyatnykh, A.A., Pikall, H., Garrido, R.: A&A **341**, 151 (1999)
2. García Hernández, A., Moya, A., Michel, E., et al.: A&A**506**, 79 (2009)
3. García Hernández, A., Moya, A., Michel, E., et al.: In: 2nd *CoRoT* symposium proceedings, p. 191 (2012)
4. Handler, G., Pikall, H., O'Donoghue, D., et al.: MNRAS **286**, 303 (1997)
5. Uytterhoeven, K., Moya, A., Grigahcène, A., et al.: A&A **534**, A125+ (2011)
6. Tassoul, M.: ApJS **43**, 469 (1980)
7. Ulrich, R.K.: ApJ **162**, 993 (1970)

Oscillation Frequency Spectra at Rapid Rotation

Fernando Pena, Robert Deupree, and Wilfried Beslin

Abstract We have computed pulsation frequencies for mid order p-modes for 2D models of rapidly rotating stars. Tracking down the evolution of a particular mode frequency as a function of rotational velocity from zero to breakup speeds allows us to identify the modes at fast rotation, and potentially one can use this relationship as a discriminant for mode identification. We also address the question of how reliably we can obtain the large separation from the computed spectrum, and what happens to the small separation as a function of rotation.

1 Introduction

In order to understand the observed frequency spectrum of a star and its link with the stellar structure, asteroseismology relies on theoretical models. Most of the analysis to date has been done by assuming that stars do not rotate, or do so slowly. Theoretical predictions can achieve great precision when the star is effectively a slow rotator, like our own Sun. However, as the structure deviates from a sphere, correlating the theoretical and observed spectra becomes non-trivial. Rotation makes the analysis difficult as it requires that models of the stellar structure and oscillations account for it.

Deupree and Beslin [5] addressed the issue of rotational frequency splittings in non-axisymmetric modes, finding that, for rotation rates as low as 20% of breakup for a 10 $M_\odot$ ZAMS, the separations in frequency (with respect to, say, the large separation) become large and non-linear. These properties greatly complicate mode identification, as frequencies of different radial order modes overlap each other.

F. Pena (✉) · R. Deupree · W. Beslin
Institute for Computational Astrophysics, Saint Mary's University, Halifax, Nova Scotia, B3H 3C3, Canada
e-mail: fpena@ap.smu.ca

Pena, F. et al.: *Oscillation Frequency Spectra at Rapid Rotation.*
Astrophys Space Sci Proc. **31**, 67–70 (2013)
DOI 10.1007/978-3-642-29630-7_13,

In this work we investigate how *uniform* rotation affects the axisymmetric low-to-high-order modes (all the way up to the asymptotic regime) of a 2.25 $M_{\odot}$ ZAMS star. In order to obtain the frequencies and eigenfunctions we use two computational codes. ROTORC [3, 4] is a 2D stellar structure code that allows both uniform and differential rotation. NRO [2] is a linear adiabatic 2D pulsation code that can also include both uniform and differential rotation.

2 Results

2.1 Accuracy Tests

2.1.1 Mode Labeling

For stars that depart significantly from zero rotation, the quantum numbers n and l are no longer good numbers to label the modes [1]. The radial quantum number n usually measures the number of nodes along the radius in a spherical star. When rotation becomes sufficiently large (deforming the spherical shape) the number of radial nodes becomes a function of latitude [2]. A similar problem occurs when using the quantum number l. In simplified terms NRO calculates the latitudinal variations as an expansion of associated Legendre polynomials (or basis functions, [2]). Because a particular mode may require several basis functions, the relationship between l and the number of latitudinal nodes is broken.

We have decided to label every mode at different rotation in the same way as Lignieres [6] did, by relating them to n and l of the non-rotating star. We do that by identifying all the modes we are interested to follow, then increase rotation by 20 km s^{-1}, label the new modes with respect to the ones for the current rotation rate and hence for the non-rotating star. In general the frequency of a particular axisymmetric mode decreases monotonically with increasing rotation [7]. Some eigenfunctions are more affected by rotation than others, so it is not such a difficult task to relate a mode to their analogous at a previous velocity step, for which we already assigned the effective quantum numbers n_0 and l_0. We also follow the eigenfunction evolution, regardless of crossings of same parity modes in the frequency space. For example, in an échelle diagram we follow the branch of eigenfunctions that at zero rotation are $l_0 = 0$, even if they cross the branch of eigenmodes $l_0 = 2$ at some radial order. We identify these branches, regardless of the state of rotation, *by the shape of their eigenfunctions*. Near the crossing point (in frequency space), the eigenmodes of both branches may be deformed, and, depending of how close they are, they may look somewhat alike. After the crossing they go back to their previous shapes. The definition of a family of modes due to the shape of the radial displacement at the surface is merely a labeling device and does not affect the value of the frequency or the eigenfunction.

2.1.2 Frequency Relative Errors

Modes that are symmetric or antisymmetric about the equator (even or odd) are treated separately. We include a set number of associated Legendre Polynomials, called basis functions [2]. For example, by using six basis functions for symmetric modes, we include in the solution all *even* terms of the associated Legendre polynomial expansion, up to degree 10 (11 for anti-symmetric). Thus, using six basis functions in NRO will provide six frequencies for a nominal value of the radial quantum number n_0 for eigenfunctions of this number of associated Legendre polynomials in the expansion. A mode labeled l_0 will always need a number of basis functions whose maximum associated Legendre polynomial is equal or greater than the original non-rotating l_0. While not proven, one might expect the modes with smaller values of l_0 to be more accurately represented with this approach. Usually the faster the rotation, the more basis functions are needed.

In order to test the accuracy of our p-mode frequencies, we define the relative error $\epsilon_{n,l,m}(trunc)$ as:

$$\epsilon_{n_0,l_0,m}(trunc) = \frac{|\omega_{n_0,l_0,m}(trunc) - \omega_{n_0,l_0,m}(8BF)|}{\omega_{n_0,l_0,m}(8BF)}, \tag{1}$$

where *trunc* is the number of basis functions used for the calculation of the frequency ω, and $8BF$ indicates that the error is always relative to eight basis functions, the maximum number of basis functions in the latitudinal series, that can be used by NRO.

At a given rotation, we find that different combinations of n_0 and l_0, produce different values of $\epsilon_{n_0,l_0,m}(trunc)$, sometimes differing in several orders of magnitudes. We noticed that the error grows with both n_0 and l_0, which is expected, as the former depends on the number of points in the radial grid we use ($\approx$ 500) and therefore higher-order modes, which near the surface oscillate significantly, have less resolution in the outer layers of the star. The dependence with l_0 may show that number of basis functions used is not enough to accurately calculate higher-degree modes.

We have found that, for the present work, 6 basis functions are good enough for modes that satisfy $n_0 \leq 20$ and $l_0 \leq 3$. As rotation increases, the error $\epsilon_{n_0,l_0,m}(6BF)$ does increase, although stays below our defined tolerance of about 1% until $V = 217\,\mathrm{km\,s^{-1}}$ (70% of the breakup velocity).

2.2 *Regularity of Frequency Spacings at Rapid Rotation*

In the series of échelle diagrams in Fig. 1 we show that there is structure in the frequency spectrum of the high-order modes, in agreement with Lignieres [6]. The fact that frequencies of different l_0 follow a comb-like structure is indicative of the large separation being approximately uniform at all speeds [7]. However we also

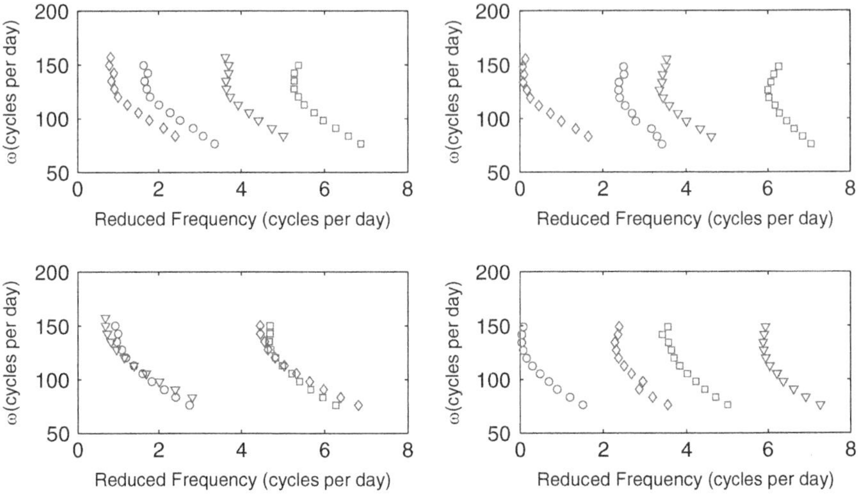

Fig. 1 Échelle diagrams for different rotational velocities of a 2.25 $M_{\odot}$ ZAMS star. *Circles* indicate modes of $l_0 = 0$, *squares* $l_0 = 1$, *diamonds* $l_0 = 2$ and *inverted triangles* $l_0 = 3$. Frequencies calculated with $8BF$. *Top left*: Initial échelle diagram at $V = 36\,\mathrm{km\,s^{-1}}$ (13% of breakup). *Top right*: $V = 87\,\mathrm{km\,s^{-1}}$ (30% breakup). *Bottom left*: $V = 131\,\mathrm{km\,s^{-1}}$ (45% breakup). *Bottom right*: $V = 195\,\mathrm{km\,s^{-1}}$ (64% breakup). Notice how the usual organization at slow rotation ($V = 36\,\mathrm{km\,s^{-1}}$) is reversed at $V = 195\,\mathrm{km\,s^{-1}}$

notice that as rotation increases different branches (different l_0) begin to separate from each other, sometimes even crossing. This shows that the small separation does change considerably with rotation, and in fact for sufficiently large rotation becomes comparable to the large separation.

Acknowledgements Fernando Pena is supported in part by an ACEnet Postdoctoral Fellowship.

References

1. Aerts, C., Christensen-Dalsgaard, J., Kurtz, D.W.: Asteroseismology. Springer, Berlin (2010)
2. Clement, M.: ApJ **116**, 57 (1998)
3. Deupree, R.: ApJ **357**, 175–187 (1990)
4. Deupree, R.: ApJ **439**, 357–364 (1995)
5. Deupree, R., Beslin, W.: ApJ **721**, 1900–1907 (2010)
6. Lignières, F., Rieutord, M., Reese, D.: A&A **455**, 607–620 (2006)
7. Lovekin, C., Deupree, R., Clement, M.: ApJ **693**, 677 (2009)

Part II
Giant Stars

Frequency Dependence of $\Delta\nu$ of Solar-Like Oscillators Investigated: Influence of HeII Ionization Zone

S. Hekker, Sarbani Basu, Y. Elsworth, and W.J. Chaplin

Abstract Oscillations in solar-like oscillators tend to follow an approximately regular pattern in which oscillation modes of a certain degree and consecutive order appear at regular intervals in frequency, i.e. the so-called large frequency separation. This is true to first order approximation for acoustic modes. However, to a second order approximation it is evident that the large frequency separation changes as a function of frequency. This frequency dependence has been seen in the Sun and in other main-sequence stars. However, from observations of giant stars, this effect seemed to be less pronounced. We investigate the difference in frequency dependence of the large frequency separation between main-sequence and giant stars using YREC evolutionary models.

1 Introduction

Stellar oscillations are an important means to study the internal structure of stars. Stars with turbulent outer layers, such as low-mass main-sequence stars, subgiants and red-giant stars, can stochastically excite oscillations in their turbulent atmospheres, so-called solar-like oscillations. These oscillations generally follow

S. Hekker (✉)
Astronomical Institute "Anton Pannekoek", University of Amsterdam, Science Park 904, 1098 HX Amsterdam, the Netherlands
e-mail: S.Hekker@uva.nl

S. Basu
Department of Astronomy, Yale University, P.O. Box 208101, New Haven CT 06520-8101, USA

Y. Elsworth · W.J. Chaplin
School of Physics and Astronomy, University of Birmingham, Edgbaston, Birmingham B15 2TT, UK

Hekker, S. et al.: *Frequency Dependence of $\Delta\nu$ of Solar-Like Oscillators Investigated: Influence of HeII Ionization Zone*. Astrophys Space Sci Proc. **31**, 73–76 (2013)
DOI 10.1007/978-3-642-29630-7_14,

a regular pattern in frequency described to reasonable approximation by the asymptotic relation derived by [12]:

$$\nu_{n,\ell} \approx \Delta\nu(n + \ell/2 + \epsilon) - \ell(\ell+1)D_0, \qquad (1)$$

with $\nu_{n,\ell}$ the frequency of an oscillation mode with radial order n and degree ℓ and $\Delta\nu$ the large frequency separation between modes of the same degree and consecutive orders. D_0 is most sensitive to deeper layers in the star and ϵ to the surface layers.

The large frequency separation can be determined for many stars in a straightforward manner, and it can be shown that this is directly proportional to the square root of the mean density of the star [9, 13]. Hence, this parameter plays an important role in studies concerning the internal structures of stars with solar-like oscillations. To a first order approximation the large frequency separation is constant over the observed frequency range. However, small deviations can occur due to for instance acoustic glitches, i.e., sudden internal property changes, but also due to other more slowly varying underlying variations, such as changing conditions close to the core.

In contrast to the clearly observed frequency dependence of $\Delta\nu$ in the Sun and main-sequence stars (e.g. Fig. 17 in [10]), first results from the *Kepler* mission [2] showed that for red giants the frequency dependence of $\Delta\nu$ is much less [6]. We investigate this difference in sensitivity of $\Delta\nu$ to the frequency range for main-sequence stars and red giants using models constructed using YREC, the Yale stellar evolution code [3]. In these models we use OPAL opacities [8] supplemented with low temperature ($\log T < 4.1$) opacities of [4] and the OPAL equation of state [11]. All nuclear reaction rates are obtained from [1], except for that of the $^{14}N(p,\gamma)^{15}O$ reaction, for which we use the rate of [5].

For more details about this study and a complete description of the different sequences of models used in this investigation, we refer to [7] and references therein.

2 Results

The observed difference in frequency dependence between main-sequence and red-giant stars could indeed be reproduced using radial oscillations computed for the YREC models. Hence, we can investigate the cause of this difference using these models.

First, we checked whether the variation of $\Delta\nu$ as a function of frequency is merely a trend, which can be resembled by a straight line with a certain slope, or due to faster variations with an amplitude dominant over the slope of a straight line. For more evolved stars the slope appears to be clearly dominant, which points to slowly varying underlying variations, such as changing conditions close to the core, to be the dominant source of the frequency dependence. For less evolved stars the amplitude of the variation of $\Delta\nu$ around a straight line is dominant, which implies a dominant contribution of acoustic glitches to the frequency dependence.

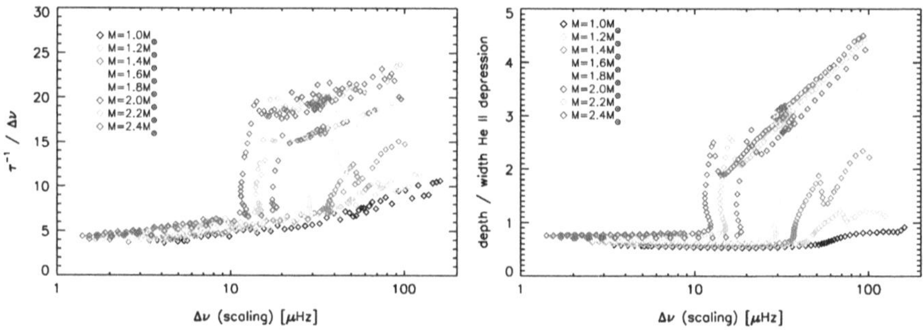

Fig. 1 *Left*: Period of the frequency modulation expressed in units of $\Delta\nu$ as a function of $\Delta\nu$. *Right*: The depth of the He II depression in Γ_1 divided by the width of the depression expressed in acoustic radius as a function of $\Delta\nu$

Interestingly, this amplitude appears to increase with increasing stellar mass and decreasing metallicity.

To investigate this further, we study the position and the shape of the depression in the first adiabatic exponent ($\Gamma_1 = (\partial \ln p/\partial \ln \rho)_s$, with p pressure, ρ density and s specific entropy) caused by the HeII ionization zone. This glitch modulates the frequencies in a sinusoidal manner with a so-called "period" inversely proportional to the acoustic depth (τ) at which the glitch is located, with

$$\tau = \int_r^R \frac{dr}{c}, \tag{2}$$

in which r indicates the radius of the glitch, R the radius of the star and c sound speed. For more evolved stars the HeII zone is located deeper in the star causing a modulation in the frequencies, i.e. dependence of $\Delta\nu$ on frequency, with a "period" much shorter (of the order of $5\Delta\nu$) than for less evolved stars, for which the "period" is much longer, up to $25\Delta\nu$ (see left panel of Fig. 1). This confirms that for evolved stars the smooth stellar structure changes are the dominant effect causing the dependence of $\Delta\nu$ on frequency, while for less evolved stars the effect from the HeII acoustic glitch is dominant.

We also investigated the influence of the shape of the depression in Γ_1 due to the HeII zone on the frequency dependence of $\Delta\nu$. We parametrize the shape of the HeII glitch by the ratio of the depth of the depression in Γ_1, i.e. its strength, to the width (in units of acoustic radius) of the depression. The narrower the width for a given depth the larger the effect the HeII depression has on the frequencies. We find that the changes in the shape of the HeII depression in Γ_1 as a function of $\Delta\nu$ (see right panel of Fig. 1) follow the same pattern as the amplitude of the variation in $\Delta\nu$ around the trend. Therefore, we conclude that the shape of the HeII depression in Γ_1 is a significant cause of the difference in the amplitudes of the variation in $\Delta\nu$. Also, the increasing amplitude of the variation in $\Delta\nu$ as a function of mass and decreasing metallicity for less evolved stars is consistently present in the shape of the HeII zone. However, there must be additional effects that play a role, because we only find qualitative agreement between the shape of the HeII depression in Γ_1 and the amplitude of the variation in $\Delta\nu$. An additional effect could be the influence of the base of the convection zone, which induces a small modulation with a period that is of the same order as the period of the HeII zone, but lower amplitude.

Acknowledgements SH acknowledges financial support from the Netherlands Organization for Scientific Research. SB acknowledges NSF grant AST-1105930. YE and WJC acknowledge financial support of UK STFC.

References

1. Adelberger, E.G., Austin, S.M., Bahcall, J.N., et al.: Rev. Mod. Phys. **70**, 1265 (1998)
2. Borucki, W.J., Koch D., Basri G., et al.: Science **327**, 977 (2010)
3. Demarque, P., Guenther, D.B., Li, L.H., et al.: AP&SS **316**, 31 (2008)
4. Ferguson, J.W., Alexander, D.R., Allard, F., et al.: ApJ **623**, 585 (2005)
5. Formicola, A., Imbriani, G., Costantini, H., et al.: Phys. Lett. B **591**, 61 (2004)
6. Hekker, S., Elsworth, Y., De Ridder, J., et al.: A&A **525**, A131 (2011)
7. Hekker, S., Basu, S., Elworth, Y., Chaplin, W.J.: (2011a) arXiv: 1109.2595
8. Iglesias, C.A., Rogers, F.J.: ApJ **464**, 943 (1996)
9. Kjeldsen, H., Bedding, T.R.: A&A **293**, 87 (1995)
10. Mathur, S., García, R.A., Régulo, C., et al.: A&A **511**, A46 (2010)
11. Rogers, F.J., Nayfonov, A.: ApJ **576**, 1064 (2002)
12. Tassoul, M.: ApJS **43**, 469 (1980)
13. Ulrich, R.K.: ApJ **306**, 37 (1986)

Some Thoughts About the Blazhko Effect for RR Lyrae Variable Pulsations

Arthur N. Cox

Abstract The amplitude variations in a few RR Lyrae stars seem possible as the beating of the main pulsation mode (a radial variation) and a nonradial $l = 1$ mode of high radial order with a period very close to the main mode. My report of this beating has been given at the Victoria (1992) and Santa Fe (2009) stellar pulsation conferences, with the first of these suggesting the radial ($l = 0$) and the $l = 1$ g4 modes occurring and the radial and the g12 modes not actually predicted to be both self excited. Recent studies show that a nonradial mode might depend on the interior structure, both based on the material opacity an its dependence on temperature and density. Current linear theory for pulsation calculations using my much improved pulsation code show that two beating modes (one radial and another nonradial) may exist for an explanation of those few pulsating RR Lyrae, amplitude varying variable stars.

1 Background

This report on the Blazhko effect is the third in a series for these stellar pulsation conferences, starting at the meeting in Victoria, Canada in 1992.

It is interesting that I did not start to work on variable stars when I was a graduate student at Indiana University. Actually I was asked to help Professor John Irwin in his summer visit to the large telescope in Pretoria in the Union of South Africa for the year 1950. We met Professor Bart Bok from Harvard University also visiting in South Africa that summer and later in Capetown. I was asked to work in Capetown with a refractor (lens) telescope for the time until summer 1951. My job was to measure the brightness and colors of the stars in a galactic cluster NGC 6231 (M41),

A.N. Cox (✉)
Los Alamos National Laboratory, 30 Bikini Atoll Rd, Los Alamos, NM 87545, US
e-mail: anc@lanl.gov

Cox, A.N.: *Some Thoughts About the Blazhko Effect for RR Lyrae Variable Pulsations.* 77
Astrophys Space Sci Proc. **31**, 77–80 (2013)
DOI 10.1007/978-3-642-29630-7_15,

Table 1 Periods and growth rates for the present model

Mode	Periods (day)	Growth rates (gain per period)
Fundamental	0.609	0.0193
1 overtone	0.453	0.0550
2 overtone	0.358	−0.00914
3 overtone	0.289	−0.1249
4 overtone	0.240	−0.2147
5 overtone	0.206	−0.2508
6 overtone	0.180	−0.2383
7 overtone	0.160	−0.2262
8 overtone	0.144	−0.2242
9 overtone	0.130	−0.2050

and the work was used by astronomers like Allen Sandage at Caltech to study the evolution of stars, especially those all close together in galactic clusters.

When I returned to Indiana University, there were half a dozen new graduate students including John Cox working on pulsating stars with a new faculty member Marshal Wrublel. After getting his PhD, John got a job as a professor at the University of Colorado. It turned out that he came to work with me in Los Alamos for many summers, as he worked on an important book [3] for the Theory of Stellar Pulsation.

My first work on the Blazhko effect for RR Lyrae stars in 1992 found that two modes (the radial fundamental and nonradial, $l = 1$, g4) were pulsationally unstable and therefore could occur in a star and beat together [1]. I tried again for the Santa Fe conference in 2009 [2], but for this time, I did not get two pulsationally unstable modes that could beat together. The results of this early work are in Table 1 below. An important question is whether the two beating modes could both be radial, but higher order modes have pulsation amplitudes in deep layers where radiation flow can produce big damping effects. I have concentrated my attention to nonradial modes with less deep damping.

For this conference I have used an RR Lyrae 0.7 solar mass model in the first figure (Fig. 1) below. Note that the high material opacity in the layers near the stellar surface gives superadiabatic regions and two convection zones.

The next figure (Fig. 2) shows the eigensolution for the radial fundamental mode that could be beating with another (even a nonradial) mode to give the Blazhko effect.

There is another thought that there may be very small power detected at longer period for many RR Lyrae variables measured by the *Kepler* satellite that might involve the radial mode overtone nine that has a beating period twice that of the second overtone. The periods of modes 1–9 are given in Table 2 below. Further research is being done now at Budapest to see if this thought may help in the interpretation of observations for many RR Lyrae variables that show the Blazhko effect.

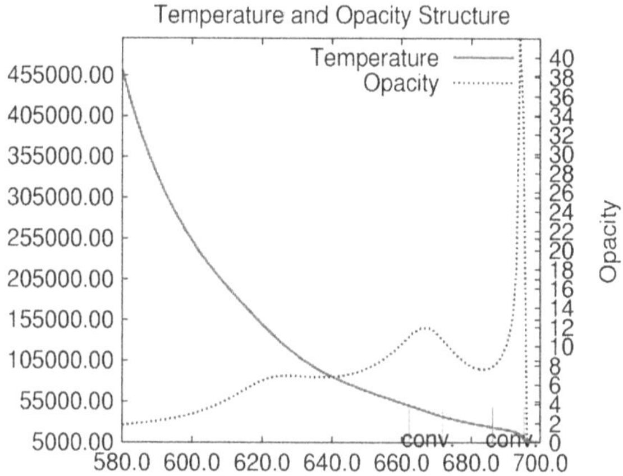

Fig. 1 Temperature (K) and opacity (cm^2/g) structure vs. mass shell number for 0.7 solar mass RR Lyr model with T_{eff} = 6,750 K, Y = 0.28, and Z = 0.001

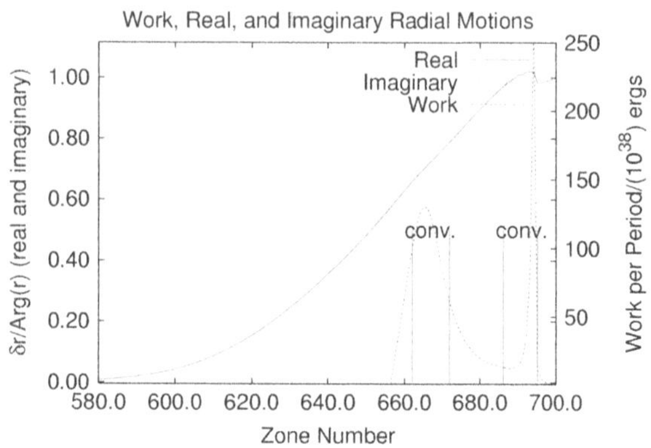

Fig. 2 Work per period (10^{38} erg), and real and imaginary eigenfunction radial amplitude vs. mass shell number for 0.7 solar mass RR Lyr model with T_{eff} = 6,750 K, Y = 0.28, and Z = 0.001

Table 2 Mass Shells of Nodes for the Overtones in Bono (private communication, 1992) 0.7 Solar Mass Model, 6,650 K, Y = 0.28, Z = 0.001

Mode	Mass shell number
1 overtone	635
2 overtone	613, 655
3 overtone	600, 631, 666
4 overtone	592, 617, 643, 673
5 overtone	586, 608, 629, 651, 678
6 overtone	581, 601, 619, 637, 658, 682
7 overtone	577, 596, 612, 627, 644, 663, 684
8 overtone	574, 591, 606, 620, 634, 649, 666, 686
9 overtone	571, 587, 601, 614, 626, 639, 653, 668, 688

The nodes in the eigenfunctions for the overtones given below are actually in the surface layers for most of the mass shell numbers given in the two figures. For these overtone modes there is always a bit of kappa effect pulsation driving where the opacity is largest, but mostly the kappa effect is strongly damping to give nonpulsating for all modes except the fundamental and the first overtone.

References

1. Cox, A.N.: New perspectives on stellar pulsation and pulsating variable stars. In: Nemec, J.M., Matthews, J.M. (eds.) Proceedings of IAU Colloquium No. 139, p. 409. Cambridge University Press, Cambridge (1992)
2. Cox, A.N.: Stellar pulsation: challenges for theory and observation. In: AIP Conference Proceedings, **1170**, Santa Fe, NM, 31 May – 5 June 2009, p. 276 (2009)
3. Cox, J.P.: Theory of Stellar Pulsation. Princeton University Press, Princeton (1980)

Period Doubling in *Kepler* RR Lyrae Stars

R. Szabó, Z. Kolláth, L. Molnár, K. Kolenberg, D.W. Kurtz, and WG#13 members

Abstract The origin of the conspicuous amplitude and phase modulation of the RR Lyrae pulsation—known as the Blazhko effect—is still a mystery more than 100 years since its discovery. With the help of the *Kepler* space telescope we have revealed a new and unexpected phenomenon: period doubling in RR Lyr—the eponym and prototype of its class—as well as in other *Kepler* Blazhko RR Lyrae stars. We have found that period doubling is directly connected to the Blazhko modulation. Furthermore, with hydrodynamic model calculations we have succeeded in reproducing the period doubling and proved that the root cause of this effect is a high order resonance (9:2) between the fundamental mode and the ninth radial overtone, which is a strange mode. We discuss the implications of these recent findings on our understanding of the century-old Blazhko problem.

1 Period Doubling in Pulsating Variable Stars

Period doubling (PD) bifurcation is a well-known dynamical effect. In the parlance of dynamical systems a new limit cycle emerges from an existing limit cycle with a period twice as long as the old one. In the case of a pulsating star we observe

R. Szabó (✉) · Z. Kolláth · L. Molnár
Konkoly Observatory of the Hungarian Academy of Sciences, H-1121 Budapest, Konkoly Thege Miklós út 15-17., Hungary
e-mail: rszabo@konkoly.hu

K. Kolenberg
Harvard-Smithsonian Center for Astrophysics, 60 Garden St., Cambridge, MA 02138, USA, Instituut voor Sterrenkunde, Celestijnenlaan 200D, 3001 Heverlee, Belgium

D. Kurtz
Jeremiah Horrocks Institute, University of Central Lancashire, Preston PR1 2HE, UK

WG#13 members
http://astro.phys.au.dk/KASC/

Szabó, R. et al.: *Period Doubling in* Kepler *RR Lyrae Stars*.
Astrophys Space Sci Proc. **31**, 81–84 (2013)
DOI 10.1007/978-3-642-29630-7_16,

alternating cycles in the time domain. In the frequency domain PD manifests itself as half-integer frequencies (the f/2 subharmonic and its odd integer multiples). In stellar astrophysics the heyday of period doubling occurred more than two decades ago when it was discovered in one-dimensional hydrodynamic models of Cepheids [1,2,11] and Type II Cepheids [4,9] by J. R. Buchler and his collaborators. The cause of PD was found to be a low-order resonance between the fundamental mode and a low radial overtone. In the hydrodynamic model calculations mentioned above, it is usually the 3:2 or the 5:2 resonances that act [2]. One of the main reasons to study the phenomenon is that a star can go from regular pulsations to chaos through a PD bifurcation cascade (known as a Feigenbaum cascade). Buchler et al. [2] proved that the characteristic light variations of RV Tauri stars, which show alternating deep and shallow minima, can be interpreted as a result of deterministic chaos of low dimension.

Even a Mira star, (R Cyg), was observed to show this phenomenon [6], and quite recently BL Herculis stars with PD were discovered in the OGLE-III data [14], confirming earlier theoretical predictions. RR Lyr stars, on the other hand, have been thought to pulsate quite regularly without low order resonances. This belief was based on decades-long observations and hydrodynamic models, with the only disturbing fact being the perplexing presence of amplitude and phase modulation (known as the Blazhko effect) in an increasing number of these stars. Therefore, the discovery of PD in the *Kepler* data was a surprise, forcing us to overhaul existing models and theories of RR Lyr pulsation.

2 Discovery of Period Doubling in RR Lyrae Stars

Kepler is a NASA Discovery mission to find Earth-like planets in the habitable zones of solar-like stars using the transit method [2]. It provides incredibly high-precision, quasi-continuous observations of a 115 $\deg^2$ swath of the sky. The currently known sample of RR Lyr stars in the *Kepler* field consists of some 40 members, the majority of which are RRab stars, and half of which show Blazhko modulation. The field contains RR Lyrae itself, and despite of its brightness, hence heavy saturation on the CCD, we managed to get extremely precise photometry for this important target with *Kepler* [16].

We found PD in the first release of data for RR Lyrae itself (KIC 7198959) [7], then subsequently in two other *Kepler* Blazhko RR Lyr stars: V808 Cyg (Fig. 1) and V355 Lyr (KIC 4484128 and KIC 7505345, respectively), both of which are much fainter than RR Lyrae itself [16]. The strength of the PD is variable; at the strongest phase the difference between subsequent maxima can be as large as 0.1 magnitude in RR Lyrae. We found that PD is stronger in certain Blazhko phases. There are hints in another four modulated stars of half-integer frequencies [16], which means that at least half of the modulated RRab stars show PD as well. After monitoring our RR Lyr star sample with *Kepler* for years we expect to see PD in more stars.

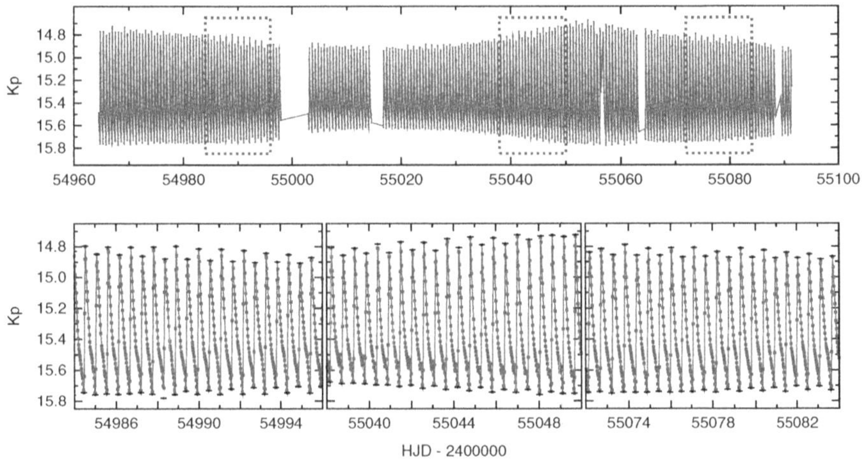

Fig. 1 *Top*: A *Kepler* light curve of the Blazhko star V808 Cygni (KIC 4484128) in quarters Q1 and Q2 showing 133 d of observations. Gaps in the data are due to safe mode events of the spacecraft and planned data download periods. The Blazhko cycle is around 90 d. *Bottom*: Magnification of three 12-d sections of the light curve highlighted by rectangles in the upper panel showing the characteristic period doubling behaviour. Polynomial fitting of the maxima and minima are also plotted for better visibility

Interestingly, non-Blazhko stars do not show PD down to the precision of the *Kepler* measurements [16], [12]. Despite close monitoring of RR Lyr stars it was not possible to detect PD previously, partly because consecutive cycles can rarely be observed from one geographical location, while the usually low amplitude of the phenomenon and its non-stationary nature also add to the difficulties.

3 Period Doubling and the Blazhko Effect

The PD phenomenon is intimately connected to the Blazhko cycle. Therefore by studying it, we may gain new insights into the intricacies of the Blazhko effect. Importantly, we succeeded in reproducing PD in hydrodynamic models [16], and unambiguously traced its cause back to a 9:2 resonance between the fundamental mode and the ninth overtone, which is a strange mode [8, 10].

Based on resonant amplitude equations accounting for the 9:2 resonance between the fundamental mode and the ninth (strange) overtone, Buchler and Kolláth [3] have found that this resonance may give rise not to only period doubled solutions, but irregularly (chaotic) modulated solutions as well. This is important, since recent observations show that the Blazhko effect is not a clockwork precision process: both long-term and cycle-to- cycle variations are frequently found in the modulation (see, e.g., [5]). Further investigations should clarify whether these amplitude equations can describe state-of-the-art hydrodynamic model calculations and, ultimately, real

RR Lyr stars. If that turns out to be the case, then this elegant prediction may be the long-sought explanation of the mysterious Blazhko effect. Additional complicating effects can also contribute to the whole picture, such as resonances involving nonradial modes [13], three-mode resonances [10] and magneto-hydrodynamic dynamo-like processes [15]. Beyond doubt, the discovery of period doubling in the *Kepler* data has opened a whole new avenue in RR Lyr star research.

Acknowledgements We gratefully acknowledge the entire *Kepler* team, whose outstanding efforts have made these results possible. This project has been supported by the 'Lendület' program, the Hungarian OTKA grants K83790 and MB08C 81013, the European Community's Seventh Framework Programme (FP7/2007-2013) under grant agreement no. 269194. R. Szabó was supported by the János Bolyai Research Scholarship of the Hungarian Academy of Sciences.

References

1. Borucki, W.J., Koch, D., Basri, G., Batalha, N., Brown, T., et al.: Kepler planet-detection mission: introduction and first results. Science **327**, 977–980 (2010)
2. Buchler, J.R., Kolláth, Z, Serre, T., Mattei, J.: Nonlinear analysis of the light curve of the variable star R Scuti. ApJ **462**, 489–501 (1996)
3. Buchler, J.R., Kolláth, Z.: On the blazhko effect in RR Lyrae stars. ApJ **731**, 24 (2011)
4. Buchler, J.R., Kovács, G.: Period doubling bifurcations and chaos in W Virginis models. ApJL **320**, L57–L62 (1987)
5. Guggenberger, E., Kolenberg, K., Chapellier, E., Poretti, E., Szabó, R., et al.: The CoRoT star 105288363: strong cycle-to-cycle changes of the Blazhko modulation. MNRAS **415**, 1577–1589 (2011)
6. Kiss, L.L., Szathmáry, K.: Period-doubling events in the light curve of R Cygni: evidence for chaotic behaviour. A&A **390**, 585–596 (2002)
7. Kolenberg, K., Szabó, R., Kurtz, D.W., Gilliland, R.L., Christensen-Dalsgaard, J., et al.: First Kepler results on RR Lyrae stars. ApJL **713**, L198–L203 (2010)
8. Kolláth, Z., Molnár, L., Szabó, R.: Period doubling bifurcation and high-order resonances in RR Lyrae hydrodynamical models. MNRAS **414**, 1111–1118 (2011)
9. Kovács, G., Buchler, J.R.: Regular and irregular nonlinear pulsation in population II Cepheid models. ApJ **334**, 971–994 (1988)
10. Molnár, L., Kolláth, Z., Szabó, R.: Uncovering hidden modes in RR Lyrae stars. These proceedings (2011)
11. Moskalik, P., Buchler, J.R.: Classical Cepheids with RV Tauri characteristics? ApJ **366**, 300–307 (1991)
12. Nemec, J.M., Smolec, R., Benkő, J.M., Moskalik, P., Kolenberg, K., et al.: Fourier analysis of non-Blazhko ab-type RR Lyrae stars observed with the Kepler space telescope. MNRAS **417**, 1022–1053 (2011)
13. Nowakowski, R.M., Dziembowski, W.A.: Resonant excitation of nonradial modes in RR Lyr stars. AcA **51**, 5–47 (2001)
14. Smolec, R., Soszynski, I., Moskalik, P., Udalski, A., Szymanski, M.K., et al.: Discovery of period doubling in BL Herculis stars of the OGLE survey. Observations and theoretical models. MNRAS **419**, 2407 (2012)
15. Stothers, R.B.: A new explanation of the Blazhko effect in RR Lyrae stars. ApJ **652**, 643–649 (2006)
16. Szabó, R., Kolláth, Z., Molnár, L., Kolenberg, K., Kurtz, D.W., et al.: Does Kepler unveil the mystery of the Blazhko effect? First detection of period doubling in Kepler Blazhko RR Lyrae stars. MNRAS **409**, 1244–1252 (2010)

First Detection of Period Doubling in a BL Herculis Type Star: Observations and Theoretical Models

R. Smolec, I. Soszyński, P. Moskalik, A. Udalski, M.K. Szymański, M. Kubiak, G. Pietrzyński, Ł. Wyrzykowski, K. Ulaczyk, R. Poleski, S. Kozłowski, and P. Pietrukowicz

Abstract We report on the discovery of the first BL Herculis star displaying period-doubling behaviour as predicted by the hydrodynamic models of Buchler and Moskalik (ApJ 391:736, 1992). The star, with $P_0 = 2.4$ d, is located in the Galactic bulge and was discovered with OGLE-III photometry. We present new nonlinear convective models, which, together with recent evolutionary tracks, put constraints on the stellar parameters. In particular, we estimate the mass and metallicity of the object.

1 Introduction

BL Herculis stars are a subgroup of type II (or Population II) Cepheids, pulsating with periods between 1 and 4 days (see [8] for a recent review). They show a singly-periodic large-amplitude light variation. Such behaviour was qualitatively reproduced with early nonlinear radiative models, e.g. Buchler and Moskalik [1, 3]. In several of their models with periods between 2 and 2.6 days, Buchler and Moskalik [1] found period-doubling behaviour—oscillations with periodic alternations of deep and shallow minima—a phenomenon not observed in any BL Her star at that time. Here we report on the discovery of the first BL Her star clearly showing

R. Smolec (✉)
Institute of Astronomy (IfA), University of Vienna, Türkenschanzstrasse 17, A-1180 Wien, Austria
e-mail: radek.smolec@univie.ac.at

I. Soszyński · A. Udalski · M.K. Szymański · M. Kubiak · G. Pietrzyński · Ł. Wyrzykowski · K. Ulaczyk · R. Poleski · S. Kozłowski · P. Pietrukowicz
Warsaw University Observatory, Al. Ujazdowskie 4, 00-478 Warszawa, Poland

P. Moskalik
Copernicus Astronomical Centre, Bartycka 18, 00-716 Warszawa, Poland

Smolec, R. et al.: *First Detection of Period Doubling in a BL Herculis Type Star: Observations and Theoretical Models*. Astrophys Space Sci Proc. **31**, 85–89 (2013)
DOI 10.1007/978-3-642-29630-7_17,

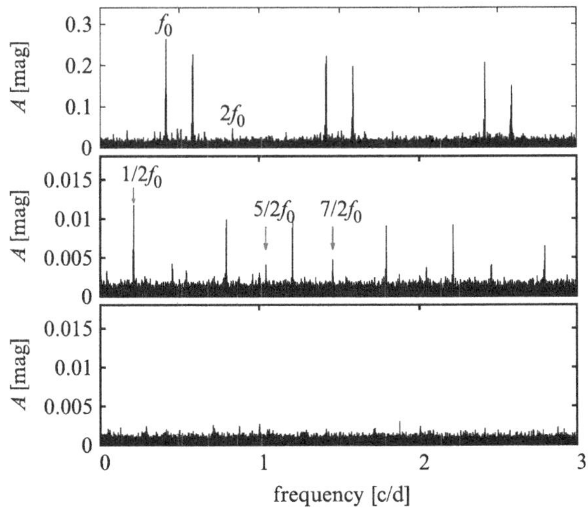

Fig. 1 Pre-whitening sequence for BLG184.7 133264. *Upper panel*: power spectrum of the original data. The spectrum is dominated by a pulsation frequency f_0 and its daily aliases. *Middle panel*: power spectrum after removing f_0 and its harmonics. The highest peak corresponds to a subharmonic frequency $\frac{1}{2}f_0$. The daily aliases are also prominent. *Lower panel*: power spectrum after removing f_0, its harmonics and its subharmonics

period-doubling behaviour. We summarize the observations and new pulsation models for this star. A more detailed analysis, as well as discussion of another BL Her star in which period-doubling behaviour is strongly suspected can be found in Smolec et al. [6].

2 Observations

The star of interest, BLG184.7 133264 is in the Galactic bulge and was discovered in data collected during the third phase of the Optical Gravitational Lensing Experiment [7]. It pulsates in the fundamental mode with a period of $P_0 = 2.4$ d. The I-band data for the star were analysed using standard consecutive pre-whitening technique. Results are presented in Fig. 1. After removing the fundamental mode frequency, f_0, and its harmonics, additional signals are visible (middle panel of Fig. 1). The dominant peak is located at $\frac{1}{2}f_0$, which is a subharmonic of the primary frequency. Other subharmonics ($\frac{5}{2}f_0$, $\frac{7}{2}f_0$ and $\frac{9}{2}f_0$) are also present. The presence of these frequencies in the power spectrum is a characteristic signature of period doubling. In the time domain it means that the light curve repeats itself after two pulsation periods, instead of one. This gives rise to strictly periodic alternations (Fig. 2).

3 Theoretical Models

Using our nonlinear convective pulsation codes [5], we have computed several model sequences in order to model the period doubling in BLG184.7 133264 and to constrain its parameters. Computational details are presented in [6]. Here we focus

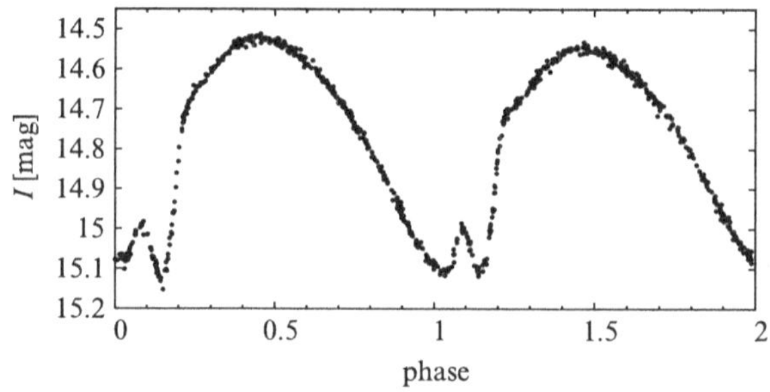

Fig. 2 Light curve of BLG184.7 133264 phased with twice the pulsation period, $2P_0$

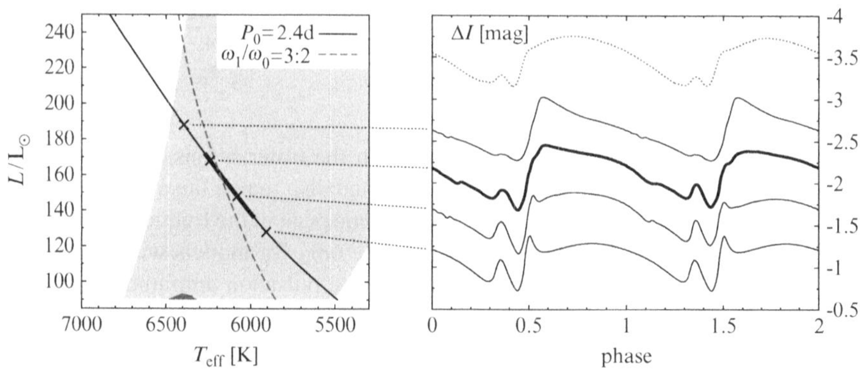

Fig. 3 The theoretical BL Her instability strip in the HR diagram for $M = 0.50\mathrm{M}_\odot$ and $Z = 0.01$ (*left panel*). Nonlinear models are computed along a line of constant period, $P_0 = 2.4$ d. The thick segment along this line indicates the period-doubling domain. Light curves for four selected models (marked with crosses) are displayed in the *right panel* (arbitrarily shifted in the vertical direction). These are to be compared with the observed light curve (represented by a Fourier fit to the data) plotted at the top

on presenting our best-fitting models and the resulting constraints on the parameters of BLG184.7 133264.

We have considered a grid of model masses ($M = 0.50\mathrm{M}_\odot$, $0.55\mathrm{M}_\odot$, $0.60\mathrm{M}_\odot$ and $0.65\mathrm{M}_\odot$) and model metallicities ($Z = 0.01$, 0.001 and 0.0001). For each (M, Z) combination, an extensive grid of linear models was computed which covers the full BL Her instability strip in the HR diagram (Fig. 3). Next, a sequence of nonlinear models was computed along a line of constant period, $P_0 = 2.4$ d (right panel of Fig. 3). Period-doubling behaviour was found over a range of luminosities. In Fig. 3 these are shown by the thick line for a particular (M, Z) combination. This domain correlates with the loci of the 3:2 half-integer resonance between the fundamental and first overtone modes, shown by the dashed line in Fig. 3. As shown by Moskalik and Buchler [2], half-integer resonances are responsible for the period-doubling behaviour. Our detailed analysis [6] confirms that the 3:2 resonance indeed causes the period doubling behaviour in the BL Her models, as Buchler and Moskalik have already shown [1].

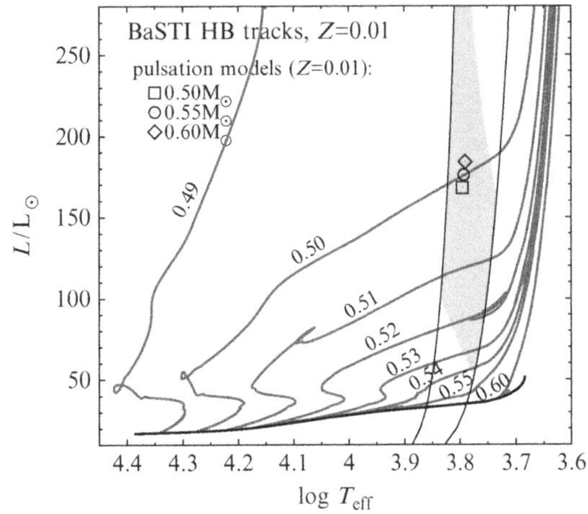

Fig. 4 The BaSTI [4] horizontal branch evolutionary tracks for $Z = 0.01$. Evolutionary tracks start at the Zero-Age Horizontal Branch (*solid*, horizontally running line). Each track is labeled with the corresponding model mass. Over-plotted are the edges of the instability strip, with the shaded area indicating the BL Her domain with fundamental mode periods between 1 d and 4 d. Pulsation models are plotted with different symbols

The model light curves were compared with the observations. Our best-fitting models closely match the pulsation amplitude and also match the amplitude of the alternations (i.e., the amplitude of the subharmonic peak in the frequency spectrum) of BLG184.7 133264. A good match is possible only for models with the highest metallicity, $Z = 0.01$. For lower metallicities, the pulsation amplitudes are lower than observed and, in addition, the amplitudes of the alternations are much larger than observed. This result is independent of the adopted values of the convective parameters, as analysed in detail in [6]. The three best-fitting models, all with $Z = 0.01$, have different masses: $M = 0.50\mathrm{M}_{\odot}$, $0.55\mathrm{M}_{\odot}$ and $0.60\mathrm{M}_{\odot}$. Based on pulsation computations alone, we cannot decide on the best model. However, since all these models fall roughly in the same place in the HR diagram (see Fig. 4), evolutionary tracks may provide further constraints.

In Fig. 4 we plot horizontal branch evolutionary tracks from the BaSTI database corresponding to $Z = 0.01$ [4]. It is clear that only our least massive model ($M = 0.50\mathrm{M}_{\odot}$) fits the evolutionary scenario. The tracks for larger masses run well beyond the instability strip. The light curve of the best model is shown in Fig. 3 (second model light curve from top, plotted with a thick line).

Acknowledgements RS is supported by the Austrian FWF grant No. AP 21205-N16. Support from Österreichische Forschungsgemeinschaft (Projekt 06/12308) is greatly acknowledged. The research leading to these results has also received funding from the European Research Council under the European Community's Seventh Framework Programme (FP7/2007-2013)/ERC grant agreement no. 246678.

References

1. Buchler, J.R., Moskalik, P.: ApJ **391**, 736 (1992)
2. Moskalik, P., Buchler, J.R.: ApJ **355**, 590 (1990)
3. Moskalik, P., Buchler, J.R.: ApJ **406**, 190 (1993)

4. Pietrinferni, A., et al.: ApJ **642**, 797 (2006)
5. Smolec, R., Moskalik, P.: AcA **58**, 233 (2008)
6. Smolec, R., et al.: MNRAS **419**, 2497 (2012)
7. Udalski, A.: AcA **53**, 291 (2003)
8. Wallerstein, G.: PASP **114**, 689 (2002)

The Projection Factor of Cepheids

N. Nardetto, J. Storm, W. Gieren, A. Fokin, and D. Mourard

Abstract The projection factor used in Baade–Wesselink methods of determining the distance of Cepheids is a crucial link between stellar physics and the cosmological distance scale. A coherent picture of this physical quantity is now available from high resolution spectroscopy and hydrodynamical modelling. Recent observations of 36 LMC Cepheids bring new constrains on the projection factor and new insights on the universality of the period–luminosity relation.

1 Introduction

For decades, Cepheids have been used to calibrate the distance scale and the Hubble constant through the well-known period–luminosity (PL) relation [4, 17–19]. Recently, using the Baade-Wesselink (BW) method to determine distances of Cepheids, Storm et al. [21], found that the K-band PL relation is nearly universal and can be applied to any host galaxy regardless of its metallicity.

The basic principle of the BW method is to compare the linear and angular size variation of a pulsating star in order to derive its distance through simple geometry. The angular diameter is either derived by interferometry [9] or using the infrared surface brightness (IRSB) relation [7]. However, when determining the linear radius

N. Nardetto (✉) · D. Mourard
Laboratoire Fizeau, UNS/OCA/CNRS UMR6525, Parc Valrose, 06108 Nice Cedex 2, France
e-mail: Nicolas.Nardetto@oca.eu

J. Storm
Leibniz-Institut für Astrophysik Potsdam, An der Sternwarte 16, D-14482 Potsdam, Germany

W. Gieren
Departamento de Astronomía, Universidad de Concepción, Casilla 160-C, Concepción, Chile

A. Fokin
Institute of Astronomy of the Russian Academy of Sciences, Moscow 109017, Russia

Nardetto, N. et al.: *The Projection Factor of Cepheids*. 91
Astrophys Space Sci Proc. **31**, 91–94 (2013)
DOI 10.1007/978-3-642-29630-7_18, © Springer-Verlag Berlin Heidelberg 2013

variation of the Cepheid by spectroscopy, one has to use a conversion projection factor from radial to pulsation velocity.

2 A Relation Between the Period and the Projection Factor

The amplitude of the radial velocity curve depends on the method used to derive the position of the spectral line profile. Two main methods are currently used: Gaussian fitting and the first moment (centroid). Only the first moment method is insensitive to the width of the spectral line (turbulence and/or stellar rotation) and is thus appropriate for comparing the dynamical structure of a Cepheid's atmosphere [2, 14]. We used this definition of the radial velocity, to derive a semi-theoretical period-projection factor (Pp) relation based on spectroscopic measurements with the *HARPS* high-resolution spectrograph.

We showed that there are three concepts involved in the decomposition of the projection factor [15]: (a) the geometric projection factor depends mainly on the limb-darkening of the star (p_0), (b) the correction due to the velocity gradient within the atmosphere (f_{grad}) and (c) the correction due to the relative motion between the optical and gas layers corresponding to the photosphere (f_{og}). For a detailed definition of these two layers, refer to Eqs. (1) and (3) of Nardetto et al. [13].

Both, (a) and (c) are deduced from geometrical and hydrodynamical models, respectively, while (b) is derived directly from high-resolution spectroscopic observations. The Fe I 4896.439 Å line is found to be the best to use in the BW methods. A coherent and consistent Pp factor relation is derived for this specific spectral line: $p = [-0.064 \pm 0.020] \log P + [1.376 \pm 0.023]$.

However, this method (first moment radial velocity of a specific spectral line) is not suitable for LMC Cepheids with spectra of very low S/N (around 10). The method currently used for distant (and thus faint) Cepheids is to cross-correlate thousands of spectral lines, derive a residual spectral line and apply the Gaussian fitting method to determine the radial velocity. Compared to a single-line measurement of the radial velocity (using the first moment method), the projection factor is actually reduced by about 5 percent relative to the cross-correlation method regardless of the period of the Cepheid [16]. This is illustrated in Fig. 1.

3 Toward LMC Cepheids

In their first paper, Storm et al. [20] directly constrained the Pp factor relation from observations. The zero point is based on HST trigonometric parallaxes of Galactic Cepheids [1, 5, 8, 12], and the slope is derived from BW distances of LMC Cepheids. All Cepheids in the LMC used by Storm et al. can be assumed to be at the same distance, leading to an extra constraint on the projection factor relation. Their relation is at 2 sigmas of the semi-theoretical relation of Nardetto et al. [15, 16]. For instance for δ Cep, they find $p = 1.41 \pm 0.05$, while the theoretical value is

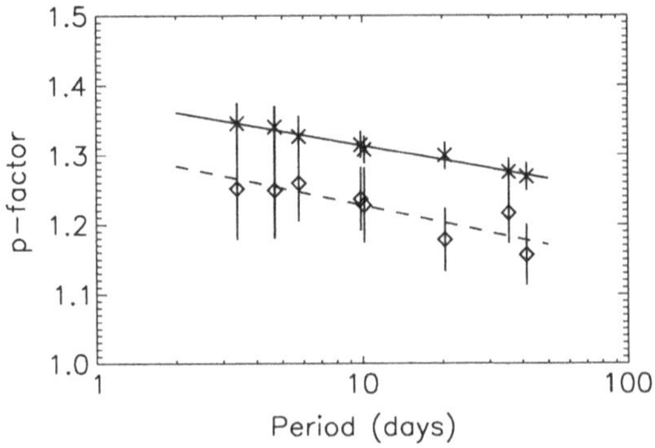

Fig. 1 The period projection factor relation (*crosses*) obtained for the very specific spectral line Fe I 4896.439 Å (low depth of about 0.1), together with the corrected relation (*diamonds*) when using the cross-correlation method. The latter relation (*diamonds*) is the one to be used to derive the BW distance of LMC Cepheids

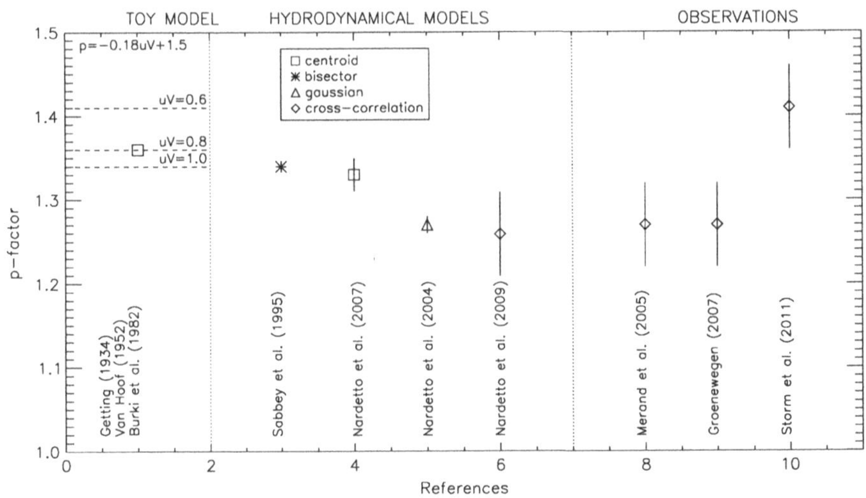

Fig. 2 The projection factor of δ Cep obtained from various approaches. (**a**) Toy modelling: various results can be easily explained when using the relationship between the geometric p-factor and limb-darkening [14]. (**b**) Hydrodynamical modelling. (**c**) Observations: the p-factor is fitted instead of the distance. The distance is derived from the HST parallax. The method used to derive the radial velocity is also critical. Consistency is obtained for MW Cepheids [6,8,13,16], but recent results from LMC Cepheids are puzzling [20]

$p = 1.25 \pm 0.05$ [16]. In Fig. 2, we show the projection factor obtained for δ Cep ($P = 5.36$ d) from various methods.

The effect of the physical nature of LMC Cepheids on the slope of the Pp relation is not excluded and should be investigated. The metallicity of Cepheids has an impact on the κ-mechanism that drives the pulsation through an opacity process. As a consequence, the amplitude of the pulsation velocity of a star is affected by the metallicity. The lower the metallicity, the larger the amplitude of the pulsation velocity [10, 11]. In principle, if the amplitude of the pulsation velocity is larger, the radial velocity should be proportionally larger and the projection factor $p = \frac{V_{\rm puls}}{V_{\rm rad}}$ should not change. The key point is then to know whether the velocity gradient in the atmosphere of the star changes with metallicity in such a way as to change the projection factor for a given spectral line. To answer this question, we consider a hydrodynamical model of δ Cep in the Milky Way [13] based on the code by Fokin et al. [3], and compare it to a consistent LMC model of the same star. We find that the projection factor is about the same for Galactic and LMC models (at the 1.5% level). This means that even if the amplitudes of the radial and pulsation velocities are larger for the model with LMC metallicity, the structure of the atmosphere is relatively unchanged. The velocity gradient is larger in absolute terms, but is unchanged relative to the velocity amplitudes.

In conclusion, the p-factor value appropriate for a Cepheid of a given period should not depend on its metallicity, and the large discrepancy obtained for the Pp relations (observation versus theory), especially for short-period Cepheids, is not explained by a difference of metallicity.

Acknowledgements WG and GP gratefully acknowledge financial support for this work from the Chilean Center for Astrophysics FONDAP 15010003, and from the BASAL Centro de Astrofisica y Tecnologias Afines (CATA) PFB-06/2007

References

1. Benedict, G.F., et al.: ApJ **133**, 1810 (2007)
2. Burki, G., Mayor, M., Benz, W.: A&A **109**, 258 (1982)
3. Fokin, A., et al.: A&A **426**, 687 (2004)
4. Freedman, W.L., Madore, B.F.: ARA&A **48**, 673 (2010)
5. Fouqué, P., Arriagada, P., Storm, J.: A&A **518**, 51(2007)
6. Getting, I.A.: MNRAS **95**, 141(1935)
7. Gieren W.P., et al.: ApJ **627**, 224 (2005)
8. Groenewegen, M.A.T.: A&A **474**, 975 (2007)
9. Kervella, P., et al.: A&A **416**, 941(2004)
10. Klagyivik, P., Szabados, L.: AN **328**, 825 (2007)
11. Klagyivik, P., Szabados, L.: A&A **504**, 959 (2009)
12. Mérand, A., et al.: A&A **447**, 783 (2005)
13. Nardetto, N., et al.: A&A **428**, 131 (2004)
14. Nardetto, N., et al.: A&A **453**, 309 (2006)
15. Nardetto, N., et al.: A&A **471**, 661 (2007)
16. Nardetto, N., et al.: A&A **502**, 951 (2009)
17. Riess, A.G., et al.: ApJ **699**, 539 (2009a)
18. Riess, A.G., et al.: ApJ **183**, 109 (2009b)
19. Riess, A.G., et al.: ApJ **730**, 119 (2010)
20. Storm, J., Gieren, W., Fouqué, P.: A&A **534**, 94 (2011a)
21. Storm, J., Gieren, W., Fouqué, P.: A&A **534**, 95 (2011b)

New Links Between Pulsation and Stellar History

Nancy Remage Evans

Abstract New instrumentation is providing new insights into intermediate mass pulsating Cepheids, particularly about their formation and history. Three approaches are discussed, using space (Hubble and Chandra) and ground-based studies (radial velocities). First, we are conducting a survey of Cepheids with the Hubble Space Telescope Wide Field Camera 3 (WFC3) to identify possible resolved companions (for example Eta Aql) and thus provide constraints on star formation. Followup X-ray observations (Chandra and XMM-Newton) can confirm whether possible low mass companions are young enough to be physical companions of Cepheids. In a related study of intermediate mass stars, Chandra X-ray observations of late B stars in Tr 16 have been used to determine the fraction which have X-ray active low mass companions. Finally, the Tennessee State Automatic Spectroscopic Telescope AST and the Moscow University group have obtained velocities of a number of Cepheids. As an example, the orbit of V350 Sgr has been redetermined, providing a new level of accuracy to the orbital velocity amplitude, which is needed for mass determination.

1 Introduction

This contribution focuses on two aspects of binary Cepheids: information they provide about star formation and about stellar evolution (masses).

N.R. Evans (✉)
SAO, 60 Garden St, MS 4 Cambridge MA 02138, USA
e-mail: nevans@cfa.harvard.edu

Evans, N.R.: *New Links Between Pulsation and Stellar History*.
Astrophys Space Sci Proc. **31**, 95–98 (2013)
DOI 10.1007/978-3-642-29630-7_19,

2 Star Formation

The formation of binary/multiple systems is a very effective way to manipulate angular momentum as a cloud collapses and stars are formed. Thus the distribution of binary parameters provides "fingerprints" of star formation, although, of course, some parameters may subsequently be altered by interactions with other stars. For solar-mass stars, binary and multiple properties have been well characterized, particularly in the seminal study of Duquennoy and Mayor [1] which combined CORAVEL radial velocities with visual binaries and common proper motion pairs. It was recently updated by Raghavan et al. [2] to include recent high resolution techniques. For more massive stars binary properties are less well determined, since they are a rarer group (hence more distant) and have broader lines (hence less accurate radial velocities). Of particular interest among the binary characteristics of intermediate/high mass stars are the frequency of binary and multiple systems, the distribution of mass ratios, and the distribution of separations (including the maximum separation). Cepheids (5 $M_{\odot}$ stars) provide a number of tools for investigating binary properties, particularly using multiwavelength techniques. The next sections describe several studies which have the over-arching goal of comparing their binary properties with those of low-mass stars.

HST Survey: We are conducting a survey of 69 of the brightest/nearest Cepheids with the Hubble Space Telescope (HST) Wide Field Camera 3 (WFC3), obtaining images in filters which transform to V and I. The goal is to look for companions as close as 100–200 AU. The fields typically cover approximately 0.1 pc = 20,000 AU, a standard expectation for the maximum dimension of binaries.

HST Survey: η Aql has been known since an early IUE (International Ultraviolet Explorer Satellite) observation to have a hot companion. However no evidence of orbital motion has been found in extensive radial velocity observations. Figure 1 (left) shows the image of the Cepheid η Aql, displaying a complex point spread function (PSF). Comparison with other WFC3 images revealed the companion, resolved at about 7 o'clock. Work is in progress on PSF correction. However for this relatively bright companion, we started with a simple subtraction. When the image of the Cepheid T Mon (scaled and aligned) was subtracted (Fig. 1 right), the companion was revealed clearly.

HST Survey: Low-Mass Companions. The most difficult companions to identify are low-mass companions with a small mass ratio relative to the primary. The second goal of the HST survey is to identify possible low-mass resolved companions. Figure 2 (left) provides as an example, the V image of R Mus. Possible companions in the field have been identified. The V-(V-I) color magnitude diagram (Fig. 2 right) shows that only one star has the appropriate color–magnitude combination to be a likely companion.

X-Rays from Low-Mass Companions. Another approach to determining the frequency of low-mass companions of high/intermediate mass stars uses X-ray

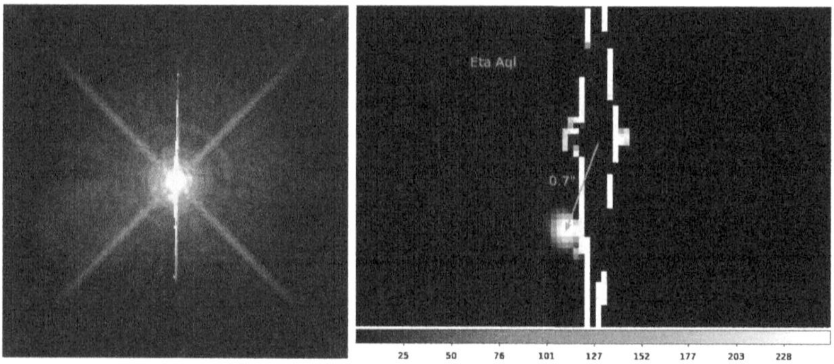

Fig. 1 *Left*: The center of the HST WFC3 V image of η Aql. The image has a log scale and is approximately $10''$ wide. *Right*: The difference image with the Cepheid T Mon subtracted from the η Aql image. The companion is clearly visible $0.7''$ from the Cepheid. Only the vertical column bleeding remains uncorrected

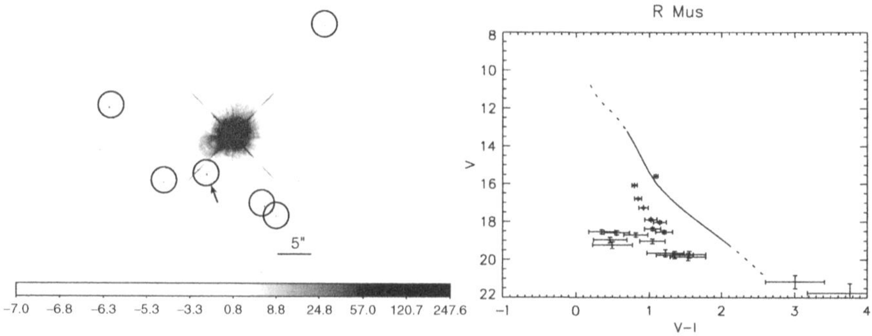

Fig. 2 *Left*: The V image of R Mus. The brighter stars among the possible companions have been *circled*, and the *arrow* indicates the probable companion from Fig. 2 (right). *Right*: The V-(V-I) diagram for faint stars in the R Mus field. The *solid line* is a ZAMS with the distance and reddening appropriate to the Cepheid. *Dotted segments* indicate stars too hot (hotter than mid-F spectral type) to produce X-rays or so cool that X-rays would be very difficult to detect

observations. Late B stars (as well as A stars) do not in general emit X-rays themselves, hence the X-rays that are occasionally found at the location of late B stars are thought to be produced by low-mass X-ray active companions (stars later than mid-F spectral type). We have developed a list of late B stars in the Carina Nebula cluster Tr 16 based on photometry and proper motions [4]. The positions of these stars were compared with the locations of X-ray source in a Chandra ACIS image of the cluster, and the results are shown in Fig. 3. From the fraction of X-ray detected stars, we find that 39 percent of late B stars have low-mass companions with masses between 1.4 and 0.5 $M_\odot$. We note that a roughly equal fraction of *high* mass companions were detected in an IUE survey of Cepheids [3], making the combined fraction of 5 $M_\odot$ stars with companions approximately three quarters.

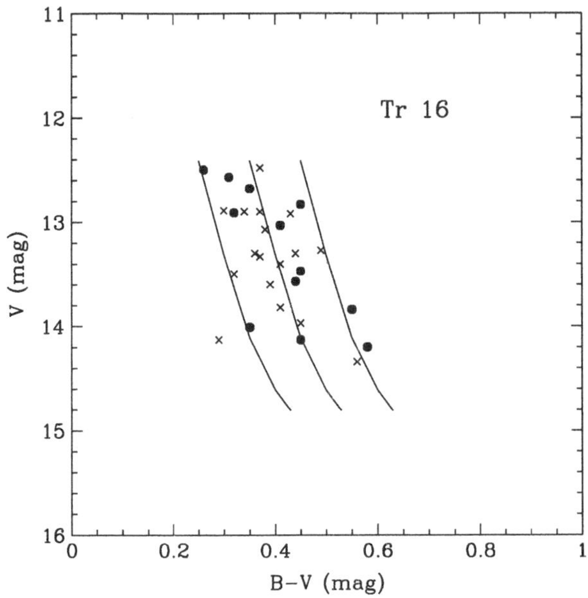

Fig. 3 The V-(B-V) colour–magnitude diagram of the late B stars in Tr 16 based on Cudworth proper motions. *Solid lines* are the ZAMS with the distance and reddening of Tr 16 (*center*) with a range in E(B-V) of ±0.1 mag (*left and right*), which is the estimated dispersion in reddening in the cluster. *Dots* are X-ray sources; *crosses* are stars not detected in X-rays. Reprinted from ApJS, 194, 13

3 Stellar Evolution: Masses

Observed mass determinations are based on spectroscopic orbits. Considerable new velocity data for Cepheids are available from sources such as the Automated Spectrographic Telescope (AST; Joel Eaton) and the Moscow University group. As an example, see the recent orbit of V350 Sgr [5]. We also have a related ongoing project to use such data sources to search for low amplitude, long period orbits of Cepheids.

Acknowledgements Financial assistance for this work was provided by Hubble grant GO-12215.01-A and the Chandra X-ray Center NASA contract NAS8-03060. Figure 1 was prepared with the help of Howard Bond and Fig. 2 with the help of Evan Tingle.

References

1. Duquennoy, A., Mayor, M.: A&A **248**, 485 (1991)
2. Raghavan, D., et al.: ApJS **190**, 1 (2010)
3. Evans, N.R.: ApJ **384**, 220 (1992)
4. Evans, N.R., et al.: ApJS **196**, 13 (2011a)
5. Evans, N.R., et al.: AJ **142**, 87 (2011b)

Evidence for Pulsation-Driven Mass Loss from δ Cephei

M. Marengo, N.R. Evans, L.D. Matthews, G. Bono, P. Barmby, D.L. Welch, M. Romaniello, K.Y.L. Su, G.G. Fazio, and D. Huelsman

Abstract We found the first direct evidence that the Cepheid class namesake, δ Cephei, is currently losing mass. These observations are based on data obtained with the *Spitzer* Space Telescope in the infrared, and with the Very Large Array in the radio. We found that δ Cephei is associated with a vast circumstellar structure, reminiscent of a bow shock. This structure is created as the wind from the star interacts with the local interstellar medium. We measure an outflow velocity of ≈ 35.5 km s^{-1} and a mass loss rate of $\approx 10^{-7}$–10^{-6} $M_{\odot}$ year^{-1}. The very low dust content of the outflow suggests that the wind is possibly pulsation-driven, rather than dust-driven as common for other classes of evolved stars.

1 Implications of Mass-Loss Processes in the Cepheid Phase

Cepheids hold the key to the cosmological distance scale. Thanks to the period–luminosity relation (Leavitt law, [9]), they are the first rung in the ladder we use to measure the size and age of the universe. They are also a benchmark for intermediate-mass stellar evolution models. Despite their importance, however, there are still outstanding puzzles in their theoretical understanding. In particular, the mass predicted by evolutionary models is significantly larger than the mass estimated by pulsation theory, or directly measured in binary systems [3, 4].

Recent calculations [2, 15] show that the Cepheid mass discrepancy can be solved by evolutionary models including convective core overshoot and mass loss. These models can dramatically lower the predicted initial mass of a Cepheid, without preventing the star from crossing the instability strip in the so-called "blue loop" characteristic of the Cepheid phase. The main difficulty with these models is that

M. Marengo (✉)
Department of Physics, Iowa State University, Ames, IA 50011, USA
e-mail: mmarengo@iastate.edu

Marengo, M. et al.: *Evidence for Pulsation-Driven Mass Loss from δ Cephei.*
Astrophys Space Sci Proc. **31**, 99–102 (2013)
DOI 10.1007/978-3-642-29630-7_20,

both overshooting and mass loss rate need to be included as free parameters, rather than independently derived from stellar physics by first principles.

Our limited understanding of stellar convection is unlikely to improve in the short term, but mass loss can be directly probed with observations. Understanding Cepheid mass loss is important not only because it affects their evolution, but also for its effect on the distance scale. Standard candles can be effectively used only if their apparent luminosity can be precisely measured. Mass loss, which surrounds the star with circumstellar material, is a source of both visible extinction and infrared excess. Both factors introduce noise in the Cepheid period–luminosity relationship, and should be taken into account if we want to reach our goal of 2% systematic accuracy in the determination of the Hubble constant [6].

While some Cepheids are known to be associated with reflection nebulæ (e.g. RS Pup), it has been difficult to prove that the circumstellar material is the result of a stellar wind. Intermediate-mass stars have a relatively short life. Such Cepheids tend to be found in regions where they are formed and where the interstellar medium (ISM) is dense. It is therefore possible that the excess emission measured in the infrared [1, 13, 16] and in the UV [5] may originate in the nearby ISM. Recent high angular resolution observations [7, 8, 14, and references therein] have revealed the presence of compact circumstellar shells around several Cepheids. The proximity of these shells to the pulsating stellar photosphere (as close as a 2–3 stellar radii) may indicate the existence of some mass loss mechanism triggered by shocks associated with stellar pulsation. Direct evidence of a large-scale mass loss process capable of explaining the Cepheid mass discrepancy has, however, eluded all observational efforts.

Our *Spitzer* observations of the Cepheid namesake δ Cephei [11] may, however, have just filled this gap. We found a large-scale structure which is best explained as an infrared bow shock resulting from the interaction of a strong stellar wind with the local ISM. Our VLA 21-cm line data [10] confirm this hypothesis and provide the first measurement of a Cepheid wind velocity as well as strong constraints on its current mass-loss rate.

2 The δ Cephei Nebula

Figure 1 shows our *Spitzer* images of the δ Cephei nebula. The emission at 8.0 and 24 μm is stronger between the star and its widely separated companion (HD 213307, a B7-8 main sequence star), and is enclosed within an arched structure detected at 70 μm. This structure is aligned with the space velocity of the star ($\approx$ 10.3 km s^{-1} and P.A. $\approx$ 58.3°). Flux ratios measured in different positions along the arc reveal a temperature of about 100 K. The composition, which has low PAH content with respect to the average ISM, is consistent with this structure being formed by a strong stellar wind pushing against the local ISM, as is common in stellar bow shocks.

This interpretation is confirmed by our VLA H I 21-cm line mapping of the region (Fig. 2). The data reveal a large ($\sim$ 1 pc) nebula with a head-tail morphology,

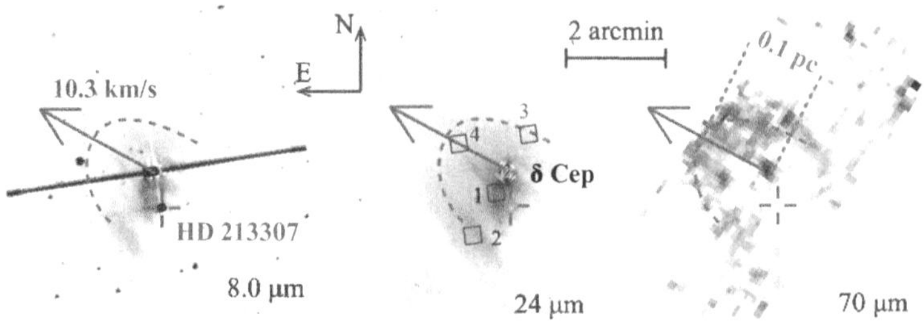

Fig. 1 *Spitzer* images of the δ Cephei nebula. The *dashed line* outlines the bright 70 μm arc. The *arrow* indicates the direction of the space velocity relative to the local ISM. The *boxes* in the middle panel show the locations where we probed the extended emission flux ratios

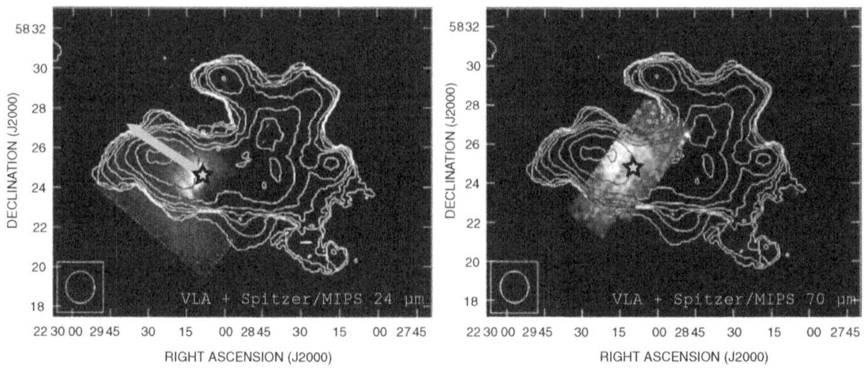

Fig. 2 VLA HI 21-cm line map of the δ Cephei nebula overlayed on our 24 and 70 μm *Spitzer* images. The location of the star is marked by the *green symbol*, and the δ Cephei space velocity is indicated by the *green arrow*

consistent with circumstellar ejecta shaped by the interaction between a stellar wind and the ISM. The bulk of the emission overlaps with the arc structure detected in the infrared, with a trailing tail in the opposite direction to the stellar space velocity. By fitting the HI line in the velocity beams not contaminated by background emission, we derived an outflow velocity of $\simeq 35.5$ km s^{-1}. This is the first actual measurement of the wind velocity for a Cepheid. It is worth noting that this velocity is significantly smaller than the expected escape velocity from the star (~ 200 km s^{-1}). Based on this measurement, the dynamical age of the structure is $\sim 10^5$ year, consistent with the expected duration of the Cepheid phase in this star.

3 Evidence for Pulsation-Driven Mass Loss

The VLA data, combined with our infrared observations, provide strong constraints on the current mass-loss rate of the star. The total flux density of the HI nebula, and detailed fitting of the 21-cm line, are consistent with a δ Cephei mass-loss rate of $\approx 10^{-6}\ M_{\odot}$ year^{-1}. This value may be considered an upper limit, since a fraction of the HI may have been swept out from the ISM, rather than having been lost from the star by the wind. An alternative estimate can be derived from the observed stand-off distance of the bow shock-like structure, together with ram pressure balance arguments and our measured wind velocity. This provides a lower limit of $\approx 10^{-7}\ M_{\odot}$ year^{-1}, by adopting a conservative estimate of the local ISM density. Models show that a mass-loss rate in this range is sufficient to solve the Cepheid mass discrepancy if sustained during the time the star is crossing the instability strip.

The total flux density detected with the VLA corresponds to a total HI mass of $\approx 0.07\ M_{\odot}$. Comparison with the dust mass derived from the infrared flux detected by *Spitzer* implies a gas-to-dust mass ratio of $\approx 2,300$. This value is significantly higher than the canonical ratio observed in dust-driven winds of evolved giant stars and in the ISM. This is consistent with the low PAH content of the δ Cephei nebula, and is further evidence that the mass-loss process acting in this star is not dust-driven.

Additional evidence is provided by the observation that the measured wind velocity is much smaller than the escape velocity. This condition requires the existence of some regulatory process capable of lifting the stellar atmosphere, rather than accelerating the flow. It also requires that most of the energy added to the wind must be in the form of momentum, rather than heat. In evolved cool stars this momentum transfer is provided by the friction of dust grains accelerated by radiation pressure. In the δ Cephei wind there is not enough dust to support this process and, due to the higher effective temperature of the star, the dust cannot form close enough to the stellar photosphere to effectively trigger this process. An alternative source for the required mechanical energy could, however, be provided by the pulsation of the star, a known source of strong shocks periodically crossing the stellar atmosphere and chromosphere [12], suggesting that this could be a pulsation-driven wind.

References

1. Barmby, P., Marengo, M., Evans, N.R., et al.: AJ **141**, 42 (2010)
2. Bono, G., Caputo, F., Castellani, V.: Mem. S. A. It. **77**, 207 (2006)
3. Caputo, F., Bono, G., Fiorentino, G., et al.: ApJ **629**, 1021 (2005)
4. Cox, A.N.: ARA&A **18**, 15 (1980)
5. Deasy, H.P.: MNRAS **231**, 673 (1988)
6. Freedman, W.L., Madore, B.F., Scowcroft, W., et al.: AJ **142**, 192 (2012)
7. Gallenne, A., Mérand, A., Kervella, P., Girard, J.H.: A&A **527**, 51 (2011)
8. Kervella, P., Mérand, A., Gallenne, A.: A&A **498**, 425 (2011)
9. Leavitt, H.S.: An. Harv. Coll. Obs., **60**, 87 (1908)
10. Matthews, L.D., Marengo, M., Evans, N,R., Bono, G.: ApJ **744**, 53 (2012)
11. Marengo, M., Evans, N.R., Barmby, P., et al.: ApJ **725**, 2392 (2010)
12. Marengo, M., Sasselov, D.D., Karovska, M., et al.: ApJ **567**, 1131 (2002)
13. McAlary, C.W., Welch, D.L.: AJ **91**, 1209 (1986)
14. Merand, A., Aufdenberg, J.P., Kervella, P., et al.: ApJ **664**, 1093 (2007)
15. Neilson, H.R., Cantiello, M., Langer, N.: A&A **529**, L9 (2011)
16. Neilson, H.R., Ngeow, C-C., Kanbur, S.M., Lester, J.B.: ApJ **692**, 81 (2009)

Multi-periodic Oscillations in Cepheids and RR Lyrae-Type Stars

P. Moskalik

Abstract Classical Cepheids and RR Lyrae-type stars are usually considered to be textbook examples of purely radial, strictly periodic pulsators. Not all the variables, however, conform to this simple picture. In this review I discuss different forms of multi-periodicity observed in Cepheids and RR Lyrae stars, including Blazhko effect and various types of radial and nonradial multi-mode oscillations.

1 Blazhko Effect

Blazhko effect is a slow, nearly periodic modulation of the pulsation amplitude and phase. It is observed in many RR Lyrae stars, both of RRab (fundamental mode) and of RRc type (overtone pulsators). The modulation period can range from a few days to over 2,500 days [44]. The phenomenon was discovered over a century ago [4,40] and was the first identified departure from the "strictly periodic" paradigm.

Blazhko effect has been detected in various stellar systems, including Magellanic Clouds [48, 49], galactic bulge [44] and several globular clusters, e.g. [30]. By no mean it is a rare phenomenon. Modulation occurs in ~50% of the field RRab stars [2, 18]. The incidence rates reported in other RR Lyrae populations are up to four times lower, perhaps because of lower quality of the available data. In most stellar systems the incidence rate of Blazhko modulation is higher in RRab stars than in RRc stars. The opposite is true only in the globular cluster Omega Centauri, where unusually high fraction (38%) of modulated RRc stars has been found [30].

Blazhko effect is usually associated with the RR Lyrae variables, but it can also occur (albeit rarely) in classical Cepheids. A modulation with a period of ~1,200 days is observed in a single-mode overtone Cepheid V473 Lyr [7]. The

P. Moskalik (✉)
Copernicus Astronomical Center, Warszawa, Poland
e-mail: pam@camk.edu.pl

Moskalik, P.: *Multi-periodic Oscillations in Cepheids and RR Lyrae-Type Stars.* Astrophys Space Sci Proc. **31**, 103–108 (2013)
DOI 10.1007/978-3-642-29630-7_21,

amplitude and phase modulation has also been recently discovered in a number of LMC double-mode Cepheids [28]. These latter variables will be discussed in Sect. 2.2.

After 100 years since its discovery, the Blazhko effect remains an unsolved mystery. Several models have been proposed to explain the phenomenon, e.g. [17, 34, 41], but all of them have failed (see excellent review by Kovács [24]). Recent discovery of the period doubling in the Blazhko RR Lyrae stars [51, 52] offers a new hope of understanding the Blazhko enigma. The period doubling is a resonant phenomenon [26]. In the RRab models it has been traced to the 9:2 resonance between the fundamental mode and the ninth radial overtone [21]. In principle, such a coupling is capable of producing not only stationary, but also modulated solutions [6]. This makes the 9:2 resonance interaction a very promising idea, that might explain the Blazhko phenomenon. But we are not there yet. So far, nonlinear RR Lyrae models have been able to reproduce the period doubling, but without the modulation [21]. It remains to be seen if *modulated* period doubling solutions can also be found.

2 Multi-mode Radial Pulsators

2.1 Double-Mode RR Lyrae-Type Star (RRd Stars)

AQ Leo, the first double-mode RR Lyrae-type variable, was identified only in 1977 [16]. Since then, double-mode RR Lyrae pulsators have been identified not only in the galactic field, but also in several globular clusters (e.g. [10, 54, 55]) and in many dwarf spheroidal galaxies, e.g. [3, 9, 11, 19, 23]. Such stars are also known in M31 [39]. By far the largest populations of RRd stars has been found in the Magellanic Clouds: 986 variables in the LMC [48] and 258 variables in the SMC [49]. For comparison, in the Galaxy we know only about 90 RRd stars in the field [56] and 80 in the bulge [44]. Typically, double-mode pulsators constitute several per cent of the RR Lyrae star population, but in some stellar systems this fraction can be as high as 20–30% [9, 54] or as low as 0.5–1% [44, 53].

In vast majority of the RRd stars the period ratio of the two excited modes, P_1/P_0, is in a narrow range of 0.742–0.748. This identifies the two modes as the first radial overtone (1O) and the radial fundamental mode (F). In most cases, the first overtone has larger amplitude than the fundamental, e.g. [48]. When plotted on the P_1/P_0 vs. P_0 plane (the so-called Petersen diagram), the RRd stars form a well defined sequence, with the period ratio being systematically higher at longer period (e.g. [48], their Fig. 4). The exact location of the star in this plot is determined mainly by its metallicity [23, 37]. In most cases, the observed period ratios correspond to [Fe/H] between -1.3 and -2.0. Only in the unique case of the galactic bulge, the RRd sequence extends down to $P_0 = 0.35$ day and $P_1/P_0 = 0.726$, which implies metallicities as high as [Fe/H] $= -0.35$ [44].

2.2 Double-Mode Cepheids

First double-mode Cepheids have been discovered more than half a century ago [35, 36]. These type of pulsators come in two basic flavours: they either pulsate in the fundamental mode and the first overtone (F/1O type, $P_1/P_0 = 0.695$–0.745) or in the first two radial overtones (1O/2O type, $P_2/P_1 = 0.79$–0.81). Recently, a third type of double-mode Cepheids has been discovered [45], with the first and the third overtones simultaneously excited (1O/3O type, $P_3/P_1 = 0.677$). So far, only two such stars have been identified.

Almost 700 double-mode Cepheids are currently known. The largest populations have been found in the LMC [25,46] (90 F/1O + 256 1O/2O + 2 1O/3O) and in the SMC [47] (59 F/1O + 215 1O/2O). For comparison, in the Galaxy we know only 24 F/1O Cepheids and 16 1O/2O Cepheids. 5 F/1O Cepheids are known in M33 [1].

The values of P_2/P_1 are almost the same for all the 1O/2O Cepheids in all stellar systems. Behaviour of P_1/P_0 is different. This period ratio becomes systematically lower as the pulsation period increases. It also differs between stellar systems, being highest in the SMC, intermediate in the LMC and lowest in the Galaxy (e.g. [28], their Fig. 1). The latter property is caused by a strong metallicity dependence of P_1/P_0.

Blazhko Effect in 1O/2O Double-Mode Cepheids

Analysis of MACHO and OGLE-II photometry of LMC variables has led to discovery of a Blazhko effect in 1O/2O Cepheids [28]. At least 20% of these double-mode pulsators display periodic modulations. Amplitudes and phases of both modes vary, with a common period, P_B, which is always longer than 700 day. The longest possible modulation period is not known, currently it is limited only by the length of the data. Modulation is stronger for the second overtone. The variations of the two amplitudes are anti-correlated: maximum amplitude of one mode always coincides with the minimum amplitude of the other mode.

The discovery of modulated 1O/2O Cepheids shows that the Blazhko effect and the double-mode pulsations are not mutually exclusive. It also imposes very strong constraints on any proposed theoretical model of the Blazhko effect. All three currently most popular models [34, 41, 50] fail to account for the properties of these stars [28]. The pattern of the modulation observed in 1O/2O Cepheids suggests that some form of energy transfer between the two modes must be involved.

2.3 Triple-Mode Cepheids

Triple-mode Cepheids are extremely rare objects. Only 8 are know so far, all have been found in the Magellanic Clouds [29, 45, 47]. They come in two different flavours: they either pulsate in the fundamental mode, the first and the second overtone (F/1O/2O type; 3 in LMC + 1 in SMC) or they pulsate in the first three radial overtones (1O/2O/3O type; 2 in LMC + 2 in SMC). All triple-mode pulsators

Table 1 F/2O double-mode RR Lyrae-type stars

Star	P_0 (day)	P_2/P_0	Reference	Star	P_0 (day)	P_2/P_0	Reference
V1127 Aql	0.3560	0.5821	[8]	V354 Lyr	0.5617	0.5862	[2]
MW Lyr	0.3977	0.5884	[17]	COROT 105288363	0.5674	0.5906	[15]
COROT 101128793	0.4719	0.5837	[38]	V350 Lyr	0.5942	0.5925	[2]
V2178 Cyg	0.4868	0.5854	[2]	KIC 7021124	0.6225	0.5931	[33]
V445 Lyr	0.5129	0.5852	[2]				

are strongly dominated by the first overtone, amplitudes of the other two modes are at least three times lower.

2.4 Secondary Modes in RRab Stars

Thanks to high quality photometry obtained with *Kepler* and COROT space telescopes, secondary periodicities with mmag amplitudes have been detected in many fundamental mode RR Lyrae variables. A group of nine objects clearly stands out (Table 1). In these stars, the ratio of the secondary and the primary periods fall in a very narrow range centered on 0.59. This is exactly the expected period ratio of the second overtone and the fundamental mode, P_2/P_0 ([2]; Smolec, private communication). The stars of Table 1 form a new group of double-mode radial pulsators: the F/2O RR Lyrae-type variables.

2.5 Multi-mode Radial Pulsations: Theory

2.5.1 Linear Theory: Modeling Pulsation Periods

Each measured period of an identified mode yields a very accurate constraint on stellar parameters. In case of multi-mode radial pulsators we have two or three such constraints, which makes these objects particularly useful for testing stellar models. In the past, analysis of double-mode Cepheids motivated the revision of stellar opacities [42]. The observed periods of double-mode variables can be used to derive metallicities of individual objects, e.g. [1, 5], and with additional input from either the observed colours or from the evolutionary tracks, the masses and luminosities of the stars or distances to stellar systems can also be determined, e.g. [12, 22]. With the triple-mode pulsators, we can constrain stellar parameters even further [27].

2.5.2 Nonlinear Models

Double-mode pulsations turned out to be very resilient to hydrodynamical modeling. The first nonlinear models displaying stable full-amplitude double-mode behaviour were computed only after time dependent turbulent convection was included into in the codes [14, 20]. The results of these calculation have recently been questioned by Smolec and Moskalik [43]. They have shown that double-mode solutions found by [14, 20] resulted from unphysical neglect of buoyancy effects in convectively stable layers of the models. At this point, the problem of reproducing the full-amplitude double-mode pulsations is far from being solved.

3 Nonradial Modes in Cepheids

Resolved low amplitude secondary frequencies have been detected in 9% of LMC first overtone Cepheids [28]. In most cases they are found very close to the frequency of the primary (radial) mode, with $|\Delta f| < 0.13$ c/d. Similar secondary periodicities have also been found in two F/1O double-mode Cepheids. Close proximity of two frequencies cannot be reproduced with the radial modes. Therefore, the secondary frequencies in these stars must correspond to *nonradial modes*. Discovery of such modes poses a challenge to the pulsation theory, which predicts that no photometrically detectable nonradial modes should be excited in Cepheid variables [32].

4 Mysterious Period Ratio of $P/P_1 \sim 0.62$

Secondary modes with puzzling period ratios in the range of 0.600–0.645 have been detected in more than 150 Magellanic Cloud Cepheids [28,46,47]. When plotted on the Petersen diagram, these variables follow two (LMC) or three (SMC) well defined parallel sequences. Secondary modes with almost the same period ratios have also been detected in 13 RR Lyrae stars [31]. In both classes of stars, such modes are found only in the first overtone and in the F/1O pulsators. The observed period ratios are incompatible with excitation of two radial modes, neither in Cepheids [13], nor in RR Lyrae stars (Smolec, private communication). Therefore, the secondary frequency must be attributed to a nonradial mode.

Kepler photometry of 4 RRc stars with $P/P_1 \sim 0.62$ revealed another intriguing feature. In all four objects, subharmonics of secondary frequencies have been detected. This means that the secondary oscillations display a *period doubling* [31].

Acknowledgements I gratefully acknowledge financial support from the conference organizers. Work on this review was also supported in part by the National Science Foundation under Grant No. NSF PHY05-51164.

References

1. Beaulieu, J.P., Buchler, J.R., Marquette, J.-B., et al.: ApJ **653**, L101–L104 (2006)
2. Benkő, J.M., Kolenberg, K., Szabó, R., et al.: MNRAS **409**, 1585–1593 (2010)
3. Bernard, E.J., Monelli, M., Gallart, C., et al.: ApJ **699**, 1742–1764 (2009)

4. Blazhko, S.: Astron. Nachr. **175**, 325–327 (1907)
5. Buchler, J.R.: ApJ **680**, 1412–1416 (2008)
6. Buchler, J.R., Kolláth, Z.: ApJ **731**, 24 (2011)
7. Burki, G., Mayor, M., Benz, W.: A&A **109**, 258–270 (1982)
8. Chadid, M., Benkő, J.M., Szabó, R., et al.: A&A **510**, A39 (2010)
9. Clementini, G., Greco, C., Held, E.V., et al.: MemSAI **77**, 249–252 (2006)
10. Corwin, T.M., Borissova, J., Stetson, P.B., et al.: AJ **135**, 1459–1473 (2008)
11. Cseresnjes, P.: A&A **375**, 909–921 (2001)
12. Dékány, I., Kovács, G., Jurcsik, J., et al.: MNRAS **386**, 521–530 (2008)
13. Dziembowski, W., Smolec, R.: AIPC **1170**, 83–85 (2009)
14. Feuchtinger, M.U.: A&A **337**, L29–L33 (1998)
15. Guggenberger, E., Kolenberg, K., Chapellier, E., et al.: MNRAS **415**, 1577–1589 (2011)
16. Jerzykiewicz, M., Wenzel, W.: AcA **27**, 35–50 (1977)
17. Jurcsik, J., Sódor, Á., Hurta, Zs., et al.: MNRAS **391**, 164–177 (2008)
18. Jurcsik, J., Sódor, Á., Szeidl, B., et al.: MNRAS **400**, 1006–1018 (2009)
19. Kinemuchi, K., Harris, H.C., Smith, H.A., et al.: AJ **136**, 1921–1939 (2008)
20. Kolláth, Z., Beaulieu, J.-P., Buchler, J.R., Yecko, P.: ApJ **502**, L55–L58 (1998)
21. Kolláth, Z., Molnár, L., Szabó, R.: MNRAS **414**, 1111–1118 (2011)
22. Kovács, G.: A&A **360**, L1–L4 (2000)
23. Kovács, G.: A&A **375**, 469–475 (2001)
24. Kovács, G.: AIPC **1170**, 261–272 (2009)
25. Marquette, J.-B., Beaulieu, J.-P., Buchler, J.R., et al.: A&A **495**, 249–256 (2009)
26. Moskalik, P., Buchler, J.R.: ApJ **355**, 590–601 (1990)
27. Moskalik, P., Dziembowski, W.A.: A&A **434**, 1077–1084 (2005)
28. Moskalik, P., Kołaczkowski, Z.: MNRAS **394**, 1649–1666 (2007)
29. Moskalik, P., Kołaczkowski, Z., Mizerski, T.: ASPC **310**, 498–501 (2004)
30. Moskalik, P., Olech, A.: CoAst **157**, 345–346 (2008)
31. Moskalik, P., Smolec, R., Kolenberg, K., et al.: These proceedings (2012)
32. Mulet-Marquis, C., Glatzel, W., Baraffe, I., Winisdoerffer, C.: A&A **465**, 937–942 (2007)
33. Nemec, J.M., Smolec, R., Benkő, J.M., et al.: MNRAS **417**, 1022–1053 (2011)
34. Nowakowski, R.M., Dziembowski, W.A.: AcA **51**, 5–47 (2001)
35. Oosterhoff, P.Th.: BAN **13**, 317–320 (1957)
36. Oosterhoff, P.Th.: BAN **13**, 320–327 (1957)
37. Popielski, B.L., Dziembowski, W.A., Cassisi, S.: AcA **50**, 491–507 (2000)
38. Poretti, E., Paparó, M., Deleuil, M., et al.: A&A **520**, A108 (2010)
39. Reyner, S., Kanbur, S.M., Ngeow, C., Morgan, C.: MNRAS **407**, 1801–1806 (2010)
40. Shapley, H.: ApJ **43**, 217–233 (1916)
41. Shibahashi, H.: ASPC **203**, 299–306 (2000)
42. Simon, N.R.: ApJ **260**, L87–L90 (1982)
43. Smolec, R., Moskalik, P.: AcA **58**, 233–261 (2010)
44. Soszyński, I., Dziembowski, W.A., Udalski, A., et al.: AcA **61**, 1–23 (2011)
45. Soszyński, I., Poleski, R., Udalski, A., et al.: AcA **58**, 153–162 (2008)
46. Soszyński, I., Poleski, R., Udalski, A., et al.: AcA **58**, 163–185 (2008)
47. Soszyński, I., Poleski, R., Udalski, A., et al.: AcA **60**, 17–39 (2010)
48. Soszyński, I., Udalski, A., Szymański, M.K., et al.: AcA **59**, 1–18 (2009)
49. Soszyński, I., Udalski, A., Szymański, M.K., et al.: AcA **60**, 165–178 (2010)
50. Stothers, R.B.: ApJ **652**, 643–649 (2006)
51. Szabó, R., Kolláth, Z., Molnár, L., et al.: MNRAS **409**, 1244–1252 (2010)
52. Szabó, R., Kolláth, Z., Molnár, L., et al.: These proceedings (2012)
53. Szczygieł, D.M., Fabrycky, D.C.: MNRAS **377**, 1263–1274 (2007)
54. Walker, A.R.: AJ **108**, 555–584 (1994)
55. Walker, A.R., Nemec, J.M.: AJ **112**, 2026–2052 (1996)
56. Wils, P.: IBVS **5955**, 1–8 (2010)

RR Lyrae Studies with *Kepler*

Katrien Kolenberg and KASC WG13 (RR Lyrae working group)

Abstract To date, about 40 known RR Lyrae stars are located in the *Kepler* field. The outstandingly high-precision data of these stars are investigated within the RR Lyrae working group as part of the *Kepler* Asteroseismic Science Consortium (KASC). I present some of the most interesting results obtained with Kepler data of RR Lyrae stars, and their implications for our understanding of RR Lyrae stars and the still mysterious Blazhko effect.

1 RR Lyrae Stars in the Kepler Field

The *Kepler* satellite was launched on March 7, 2009, and since shortly thereafter it has been staring at the same field in the sky. As a tool to detect transits of Earth-sized planets around stars, it delivers light curves of unprecedented precision (Fig. 1) of a variety of targets [2], including variable and pulsating stars [5].

RR Lyrae stars come in different flavors, and in the *Kepler* field we find radial fundamental mode pulsators, the RRab stars, many of which are modulated [1], and several radial first overtone pulsators, RRc stars. At the time of writing, about 40 RR Lyrae stars are known in the *Kepler* field with mean *Kepler* magnitudes (wide passband between 4,300 and 9,000 Å) in the range $7.9 \leq K_p \leq 17.4$.

Whereas long ($\simeq$30-min) cadence provides a good coverage of the RR Lyrae light curve, only short ($\simeq$1-min) cadence allows us to distinguish specific features in the light curves, such as a shoulder on the steep rising branch of the curve, and

K. Kolenberg (✉)
Harvard-Smithsonian Center for Astrophysics, 60 Garden Street, Cambride MA 02138, USA
Instituut voor Sterrenkunde, Celestijnenlaan 200D, B-2001 Heverlee, Belgium
e-mail: kkolenbe@cfa.harvard.edu

KASC WG13 (RR Lyrae working group)
http://astro.phys.au.dk/KASC/

Kolenberg, K. and KASC WG13 (RR Lyrae working group): *RR Lyrae Studies with* Kepler. Astrophys Space Sci Proc. **31**, 109–113 (2013)
DOI 10.1007/978-3-642-29630-7_22,

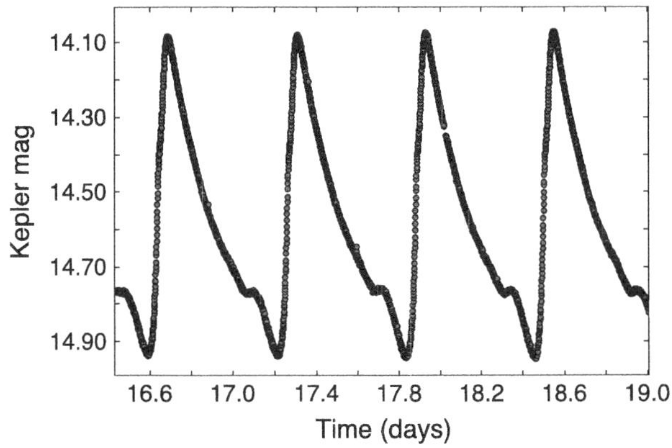

Fig. 1 Short cadence *Kepler* data of an RRab star

the bump closely before minimum light. Therefore, all of our targets are observed at least one quarter (3 months) in short cadence over the duration of the mission.

2 Some Examples and Results

2.1 Time Series Analysis

A first look at the *Kepler* data [9] already revealed that the *incidence rate* of the Blazhko effect in the *Kepler* field RR Lyraes is higher than the previously assumed 25–30% in Galactic populations. Moreover, *additional frequencies*, beyond the main pulsation mode, its harmonics and the Blazhko modulation components were detected for the first time. The half-integer frequencies associated with a period doubling effect were never before observed in RR Lyrae stars. A study based on the first quarters of all available data at the time [1] confirmed that about half of the RRab stars in our sample are Blazhko stars and modulation always occurs in both amplitude and phase. Furthermore, period doubling was found in several other Blazhko stars. In four stars, frequencies were found close to the first and/or second overtone modes, with amplitudes that seem to vary over the Blazhko cycle. Finally, two stars in our sample appear to pulsate in the fundamental and (low-amplitude) second overtone modes, thereby representing a new type of double-mode behavior [1, 13].

By a fortunate coincidence, RR Lyr itself, the prototype and by far the brightest star of the class, lies in the *Kepler* field. Thanks to a custom aperture devised by Steve Bryson (NASA Ames), the flux of this strongly saturated star could be recovered. RR Lyr offers a case study for observations of bright stars with *Kepler* [10].

One of the most challenging questions in analyzing *Kepler* data of a Blazhko star is how to correctly stitch together data from different quarters [4]. For modulated stars, it is indeed not trivial to disentangle whether variations between quarters are instrumental or stellar. Several interesting individual stars with variable Blazhko cycles are being studied in detail by working group members, e.g., [4, 6].

Interestingly, all four RRc variables known in the *Kepler* field thus far turned out to be multiperiodic, showing an additional period with a ratio to the main period of $\simeq 0.60 - 0.63$ (see [12]).

Non-modulated RR Lyrae stars allow us to study the stability of the pulsations of RR Lyrae stars and they provide a reference against which the Blazhko RR Lyrae stars can be compared [13]. From the Fourier spectra of these targets, [Fe/H] values and other physical parameters can also be derived. From this, four of our target stars turn out to be particularly metal-rich.

2.2 Theoretical Modelling

When the observed periods of the *non-modulated stars* are compared with periods computed with the Warsaw non-linear convective pulsation code better agreement is achieved assuming pulsational luminosity (L) and mass (M) values rather than the (higher) evolutionary L and M values [13].

The detection of additional frequencies in many of our RR Lyrae targets has sparked modelling efforts to explain the observations. For the period doubling these modelling efforts are discussed extensively in these proceedings by Róbert Szabó [17]. These findings even led to a new model for the Blazhko effect in the framework of radial resonances [3]. Model calculations of RR Lyrae stars suggest that the bifurcation cascade, responsible for period doubling, may evolve to chaos in these dynamical systems [14]. The possibilities and occurrences of mutual resonances between the fundamental mode and multiple overtones was also investigated in hydrodynamical models [11].

A comparison between the modulation in the *Kepler* data of the prototype RR Lyr with hydrodynamical models [15], allowed to show some limitations of the Stothers model [16]. Finally, it still remains a riddle how to explain the newly detected frequencies in RRc stars. With our current knowledge of RR Lyrae pulsation models, they can only be explained in terms of non-radial modes.

2.3 Ground-Based Follow-Up

Multicolour photometry and spectroscopy has been gathered for the majority of our targets by several working group members. Besides applications for individual stars, this will allow us to better locate our sample stars in the HR diagram and to determine their metallicities. The high-resolution spectroscopy confirms the high metallicity of our targets that had emerged from the Fourier decomposition [13]. Time-resolved spectroscopic data would also allow a more literally in-depth study of the RR Lyrae variables.

2.4 Additional Projects, and Stars!

We know that there must be many more RR Lyrae stars in the *Kepler* field, and are actively searching for them. New pulsators can be found from the *Kepler* Full Frame Image Variability Catalog [7, 8]. A few other RR Lyrae stars were already found through reclassification and some blended ones were found serendipitously (one even by an amateur through http://www.planethunters.org/!). We hope to significantly enlarge our sample of RR Lyrae stars before the end of the *Kepler* mission.

With the data from the *Kepler* mission major breakthroughs in our understanding of RR Lyrae stars have been and can/will be achieved. These data will provide material for thought and discovery for many years, possibly even decades, to come.

Acknowledgements We gratefully acknowledge the entire *Kepler* team, whose outstanding efforts have made these results possible. KK acknowledges Marie Curie Fellowship 2009-IOF-255267.

References

1. Benkő, J.M., Kolenberg, K., Szabó, R., Kurtz, D.W., Bryson, S., et al.: Flavours of variability: 29 RR Lyrae stars observed with *Kepler*. MNRAS **409**, 1585–1593 (2010)
2. Borucki, W.J., Koch, D., Basri, G., Batalha, N., Brown, T., et al.: *Kepler* Planet-detection mission: introduction and first results. Science **327**, 977–980 (2010)
3. Buchler, J.R., Kolláth, Z.: On the Blazhko Effect in RR Lyrae Stars. ApJ **731**, 24 (2011)
4. Çelik, L., Ekmekçi, F., Nemec, J., Kolenberg, K., Nemec, J., et al.: How to correctly stitch together data of a Blazhko Star? These proceedings (2011)
5. Gilliland, R.L., Brown, T.M., Christensen-Dalsgaard, J., Kjeldsen, H., Aerts, C., et al.: *Kepler* Asteroseismology program: introduction and first results. PASP **122**, 131–143 (2010)
6. Guggenberger, E., Kolenberg, K., Nemec, J., Szabó, R., Benkő, J.: Changing Blazhko modulations: KIC6186029 in comparison with other cases of non-repetitive Blazhko cycles. These proceedings (2011)
7. Kinemuchi, K., Howell, S., Still, M., Szabó, R.: Selecting new pulsating stars from the *Kepler* full frame image variability catalog! These proceedings (2011)
8. Kinemuchi, K., et al.: in preparation
9. Kolenberg, K., Szabó, R., Kurtz, D.W., Gilliland, R.L., Christensen-Dalsgaard, J., et al.: First *Kepler* results on RR Lyrae stars. ApJL **713**, L198–L203 (2010)

10. Kolenberg, K., Bryson, S., Szabó, R., Kurtz, D.W., Smolec, R., et al.: *Kepler* photometry of the prototypical Blazhko star RR Lyr: an old friend seen in a new light. MNRAS **411**, 878–890 (2011)
11. Molnár, L., Kolláth, Z., Szabó, R.: Uncovering hidden modes in RR Lyrae stars. These proceedings (2011)
12. Moskalik, P., Smolec, R., Kolenberg, K., Nemec, J., Kunder, A., Chadid, et al.: Discovery of peculiar double-mode pulsations and period doubling in *Kepler* c-type RR Lyrae stars. These proceedings (2011)
13. Nemec, J.M., Smolec, R., Benkö, J.M., Moskalik, P., Kolenberg, K., et al.: Fourier analysis of non-Blazhko ab-type RR Lyrae stars observed with the *Kepler* space telescope. MNRAS **417**, 1022–1053 (2011)
14. Plachy, E., Kolláth, Z., Molnár, L.: Detection of chaos in RR Lyrae models. These proceedings (2011)
15. Smolec, R., Moskalik, P., Kolenberg, K., Bryson, S., Cote, M.T., Morris, R.L.: Variable turbulent convection as the cause of the Blazhko effect – testing the Stothers model. MNRAS **414**, 2950–2964 (2011)
16. Stothers, R.B.: A new explanation of the Blazhko effect in RR Lyrae stars. ApJ **652**, 643–649
17. Szabó, R., Kolláth, Z., Molnár, L., Kolenberg, K., Kurtz, D.W., WG13 members: Period doubling in *Kepler* RR Lyrae stars. These proceedings (2011)

The Nature of Variability in M Supergiants: The Forgotten Type C Semiregulars

D.G. Turner, K.E. Moncrieff, C.I. Short, and R.F. Wing

Abstract The characteristics of SRC variables are summarized from a comprehensive survey of the Galactic population based on new and archival spectroscopic (Moncrieff, Ph.D. Thesis, Saint Mary's University, 2011) and spectrophotometric (White and Wing, Astrophys. J. 222:209–219, 1978) observations, and brightness estimates from Harvard plates and AAVSO visual archives, with attention on Berkeley 87 member BC Cyg. New spectroscopic observations of BC Cyg phased using a $692^{d}.8$ period, found by a new analysis tied to times of light maximum, are consistent with radial pulsation, similar to that in Cepheids. Results for other SRCs are more ambiguous, with starspot activity or supergranulation cycles constituting a complication for small-amplitude stars displaying longer term trends.

1 Introduction

Variable M supergiants constitute a "forgotten" group of long-period variable (LPV) stars, frequently overlooked because they are challenging to observe: periods of order 1–3 years, photometry difficult to calibrate, and complex spectra. The most populous LPV groups include Mira variables and types A and B semiregulars (SRA, SRB), while M supergiant variables are either semiregular (SRC) or irregular (LC).

The 1990 edition of the *General Catalogue of Variable Stars* lists 55 SRC variables and 69 LC stars, although some SRCs are probably misidentified SRDs

D.G. Turner (✉) · K.E. Moncrieff · C.I. Short
Saint Mary's University, Halifax, NS, B3H 3C3, Canada
e-mail: turner@ap.smu.ca; kathleen.moncrieff@gmail.com; cishort@ap.smu.ca

R.F. Wing
Ohio State University, Columbus, OH 43210, USA
e-mail: wing@astro.ohio-state.edu

Turner, D.G. et al.: *The Nature of Variability in M Supergiants: The Forgotten Type C Semiregulars*. Astrophys Space Sci Proc. **31**, 115–119 (2013)
DOI 10.1007/978-3-642-29630-7_23,

or LCs. By comparison, the *Catalogue* contains 460 classical Cepheids and 179 Cepheids of unknown type, an order of magnitude larger sample than variable M supergiants. Yet M supergiants may be $\sim$ 100 times more common than Cepheids in galaxies [6] and possibly more useful distance indicators because of their high luminosities ($10^{4.5-5.5} L_{\odot}$). They are also the predicted progenitors of Type II SNe.

Type C semiregulars have periods of $\sim$ 100^{d}–800^{d} originating from pulsation [7], with longer $1{,}000^{d}$–$3{,}000^{d}$ periods in some possibly representing starspot activity or supergranulation cycles [1]. Amplitudes ΔV are of order 1^{m}–6^{m} for pulsation, the larger values generally applying to cooler stars, and 1^{m}–3^{m} for longer-term trends. Spectral types range from K5 to M10 and luminosity classes from Ia to II, hotter stars (< K5) being more likely to be UU Her variables. The LC irregulars have similar spectral types, but generally smaller amplitudes, $\leq 1^{m}$, with little evidence for periodicity [5]. The group may be a catch basin for low amplitude SRCs.

Our knowledge of SRCs comes mostly from study of their light curves, the nature of their variability remaining debatable. Do most pulsate like Miras, SRAs, and SRBs [7]? Existing photometric data are few, mainly because of the long time scales of variability in the stars, but recent CCD surveys are beginning to remedy that. There is a limited amount of photoelectric data, a wealth of potential photographic data in plate archives like the Harvard collection, and almost a century of visual observations for some stars in the AAVSO archives, available on-line.

2 Light Curve Analysis

The 15 years of photoelectric photometry for α Ori and TV Gem by Wasatonic et al. [10] represent the longest sequence of precision photometry available for SRCs. More typical is the situation for BC Cyg illustrated in Fig. 1, where long-term estimates of photographic (pseudo-*B*) and visual (pseudo-*V*) brightness are available from measures made at the Harvard Plate Stacks [9] and the AAVSO archives. BC Cyg makes a useful reference star and potential Rosetta Stone for the study of SRC variables because it is also a member of the open cluster Berkeley 87 [9].

Unfortunately, there are large systematic offsets in some AAVSO visual estimates, as tested for BC Cyg by combining photographic and visual estimates into a $B - V$ color. Color changes in BC Cyg are negligible over its cycle, so the observed scatter [9] originates in the AAVSO estimates (*V* too faint). Recent AAVSO visual estimates for BC Cyg display a similar problem, namely overly large scatter amounting to $\sim 1^{m}$ for the combined observations.

Fourier techniques and O–C analyses are not always best suited to determining periods of variability in SRC variables. A better way is to track times of light maximum, established from parabolic fits to the observations, as a function of cycle count, as shown in Fig. 2. In the case of BC Cyg, a long-term period of 692.8 days is indicated. Such information is essential for linking brightness changes to changes in the star's dimensions and surface temperature.

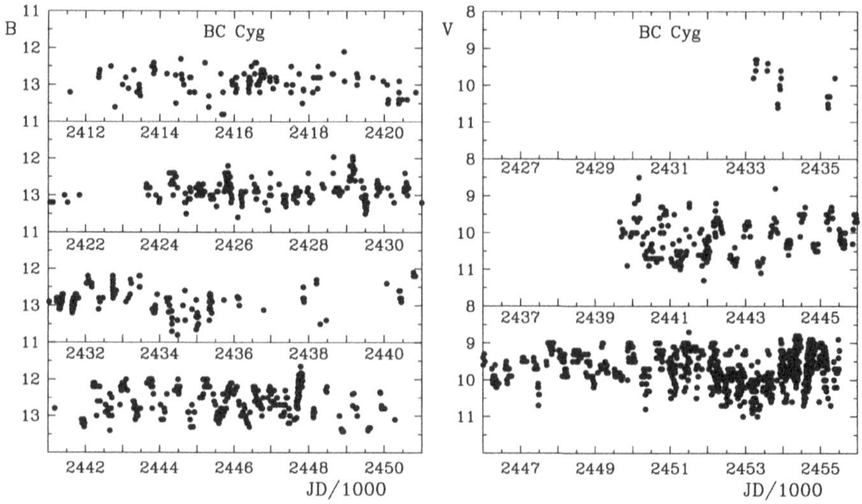

Fig. 1 Photographic (*B*, *left*) and visual (*V*, *right*) light curves for the SRC variable BC Cygni over the past century from Harvard plate estimates and AAVSO observations, respectively

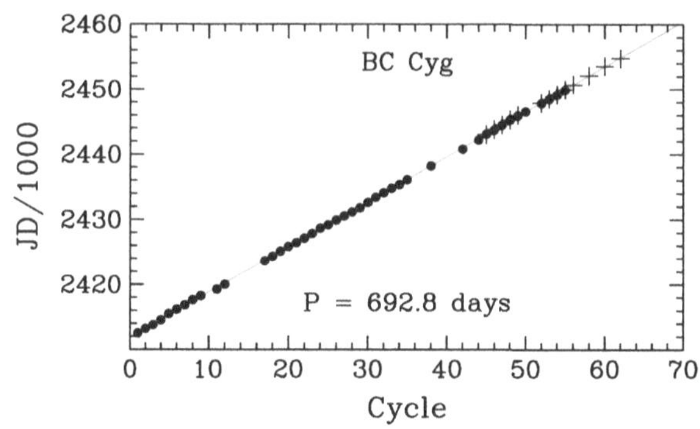

Fig. 2 Measured times of light maximum in BC Cyg from brightness estimates off photographic plates (*points*) and from AAVSO visual archives (*crosses*) versus cycle count. The line represents the long-term trend

3 Spectroscopic Information

M supergiants are difficult to classify accurately. Keenan's criteria for the temperature classification of M giants according to the strength of TiO bands in their spectra [2] works well for M supergiants. TiO band strength is also used in the White and Wing photometric temperature classification scheme [11], sensitive to ± 0.1 spectral subtype. Luminosity classification in the blue rests on the strengths of V I ($\lambda 4,391$ Å) and Y II ($\lambda 4,377$ Å) relative to lines like Fe I ($\lambda 4,382$ Å). The White and Wing scheme relies on CN strength in the red region. For BC Cyg, we

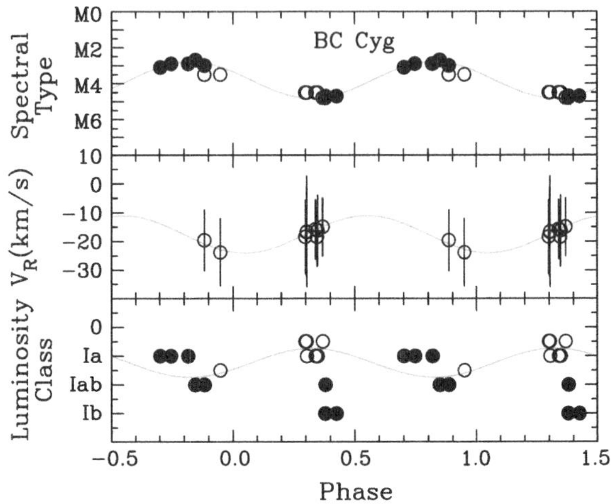

Fig. 3 Spectroscopic changes in BC Cyg versus photometric phase (spectrophotometric data: *filled circles*, spectroscopic data: *open circles*), with sine waves added to denote the trends

have used spectra at dispersions of 120 Å mm^{-1} and more often 60 Å mm^{-1} to track temperature, luminosity, and radial velocity changes in the star over the past six years, with results shown in Fig. 3. There is a small discrepancy between the spectroscopic and spectrophotometric results for luminosity variations in the star, but the radial velocity results substantiate the validity of the spectroscopic classifications.

The observations indicate that BC Cyg pulsates much like Cepheids, being hottest when it is smallest, 0.2 cycle before light maximum, and coolest when it is largest, half a cycle later, the radial velocity variations being a mirror image of the light variations with a small phase offset of +0.05 cycle. The photometric results for α Ori and TV Gem [10] show similar variations, suggesting that the dominant light variations in SRCs arise from pulsation.

Observations are in hand for another two dozen SRC variables. Some display characteristics similar to BC Cyg [3], others, like α Her and S Per [4], are more ambiguous. Such cases are still being investigated. The nature of systematic offsets for some AAVSO observers also needs investigating, since they affect light curve data for some objects, and are essential for establishing reliable ephemerides.

A Galactic calibration of the infrared PL relation for SRC variables has been used previously to determine distances to the LMC and M33 [6], but the scatter in SRC PL relations is large [3, 7, 8], affecting their potential usage. A displacement of the PL relation for cluster and association SRC variables from the Cepheid PL relation may indicate effects of significant mass loss, but that needs further evaluation.

The short period cyclical variations in brightness in most SRC variables clearly result from radial pulsation, while longer-term trends are more likely of photospheric origin. In α Ori for example, its phased angular diameter variations for a 424^d pulsation period produce results similar to BC Cyg, but not when phased to its longer 2322^d period. Here granulation and giant convection cells [1] may apply. As the predicted potential progenitors to Type II SNe, SRC variables are intriguing pulsating stars, and the secrets of their variability are just beginning to be uncovered.

References

1. Gray, D.F.: Astron. J. **135**, 1450–1458 (2008)
2. Gray, R.O., Corbally, C.J.: Stellar Spectral Classification. Princeton University Press, Princeton (2009)
3. Moncrieff, K.E.: Ph.D. Thesis, Saint Mary's University (2011)
4. Moncrieff, K.E., Turner, D.G., Short, C.I., Bennett, P.D., Griffin, R.: Odessa Astron. Publ. **23**, 86–89 (2010)
5. Percy, J.R., Terziev, E.: J. Am. Assoc. Var. Star Observ. **39**, 1–9 (2011)
6. Pierce, M.J., Jurcevic, J.S., Crabtree, D.: Monthly Notices Roy. Astron. Soc. **313**, 271–280 (2000)
7. Stothers, R.: Astrophys. J. **156**, 541(1969)
8. Turner, D.G.: Odessa Astron. Publ. **18**, 123–126 (2006)
9. Turner, D.G., Rohanizadegan, M., Berdnikov, L.N., Pastukhova, E.N.: Publ. Astron. Soc. Pacific **118**, 1533–1544 (2006)
10. Wasatonic, R.P., Guinan, E., Engle, S.: Bull. Am. Astron. Soc. **218**, 322.05 (2011)
11. White, N.M., Wing, R.F.: Astrophys. J. **222**, 209–219 (1978)

Pulsational Light Variability in Post-AGB Stars in the Milky Way Galaxy and the Magellanic Clouds

Bruce J. Hrivnak, Wenxian Lu, and Kristie A. Shaw

Abstract We recently published the results of a long-term photometric monitoring program of 12 carbon-rich post-AGB stars of F–G spectral types in the Milky Way Galaxy (MWG). These objects appear to be the direct precursors of planetary nebulae and are classified as proto-planetary nebulae (PPNe). They all vary in brightness with pulsation periods in the range of 35–155 days, with the cooler ones possessing longer periods. We have extended the study in several ways, including the addition of several oxygen-rich PPNe from the MWG and two carbon-rich PPNe from the Magellanic Clouds, and by increasing the data sets for the initial 12. The oxygen-rich objects show some clear differences; these and other results are discussed.

1 Introduction

Post-AGB stars are objects that have evolved beyond the AGB phase, and they are typically evolving horizontally across the upper part of the HR diagram at increasing temperature and constant luminosity. We are focusing our study on the subclass that can confidently be called proto-planetary nebulae (PPNe). These stars are of intermediate temperature (5,000–30,000 K), surrounded by a detached, expanding shell of gas and dust, and they are expected to become planetary nebulae within a few thousand years [6]. At present ~100 of these have been identified in the Milky Way Galaxy (MWG). Unfortunately none of these is in the *Kepler* field.

We are studying these objects because they offer the opportunity to find the properties of intermediate-mass stars while they are in this transient evolutionary phase. Pulsational light variability has the potential to allow us to determine physical

B.J. Hrivnak (✉) · W. Lu · K.A. Shaw
Department of Physics & Astronomy, Valparaiso University, Valparaiso, IN 46383 USA
e-mail: bruce.hrivnak@valpo.edu; wen.lu@valpo.edu; kristie.shaw@valpo.edu

Hrivnak, B.J. et al.: *Pulsational Light Variability in Post-AGB Stars in the Milky Way Galaxy and the Magellanic Clouds*. Astrophys Space Sci Proc. **31**, 121–125 (2013)
DOI 10.1007/978-3-642-29630-7_24,

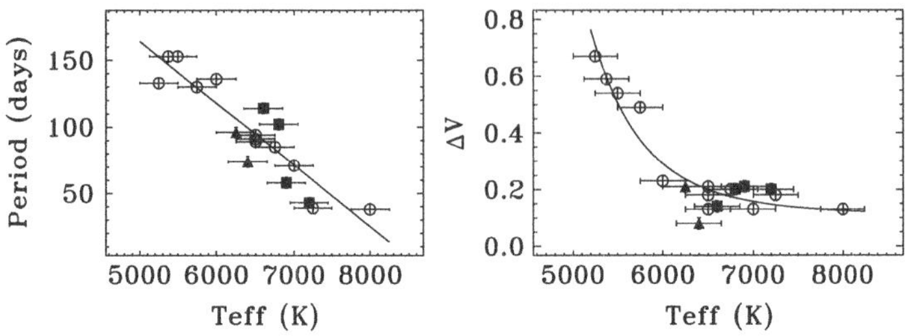

Fig. 1 Comparison of the periods of pulsation (*left*) and the peak variations (*right*) with the effective temperatures of the PPNe. The *open circles* represent the C-rich objects and the *solid lines* the best fit to these. Shown for comparison are the 2 C-rich MC PPNe (*solid triangles*) and the 4 O-rich MWG PPNe (*solid squares*) referred to in this study

properties of the stars and to gain insight into the AGB and post-AGB mass loss and evolutionary timescales.

2 Observations

Observations of ~30 PPNe and candidates have been carried out at the Valparaiso University Observatory (VUO) since 1994. The targets are relatively bright, V = 8–14 mag, and easily observable with a small and dedicated telescope. These observations provide a unique data set due to their long, continuous, multi-color (*VR*) nature. Some PPNe have also been observed by Arkhipova and collaborators [1].

3 Initial Results

We have recently completed the study of a sample of 12 carbon-rich PPNe based on 14 years of observations. They all have spectral types of F or G. We found that all of the stars varied in light due to pulsation, with periods ranging from 35 to 155 days and seasonal peak-to-peak amplitudes ranging from 0.15 to 0.60 mag in *V*, with the cooler stars having the longer periods and larger amplitudes of variation. These are shown in Fig. 1 as a function of temperature. Since the temperatures increase with time as the stars evolve, this sample demonstrates how the period and light amplitude decrease with time. Over the temperature range 5,250–8,000 K, the period decreases at a rate of ~0.05 day/K. They are all redder when fainter [4].

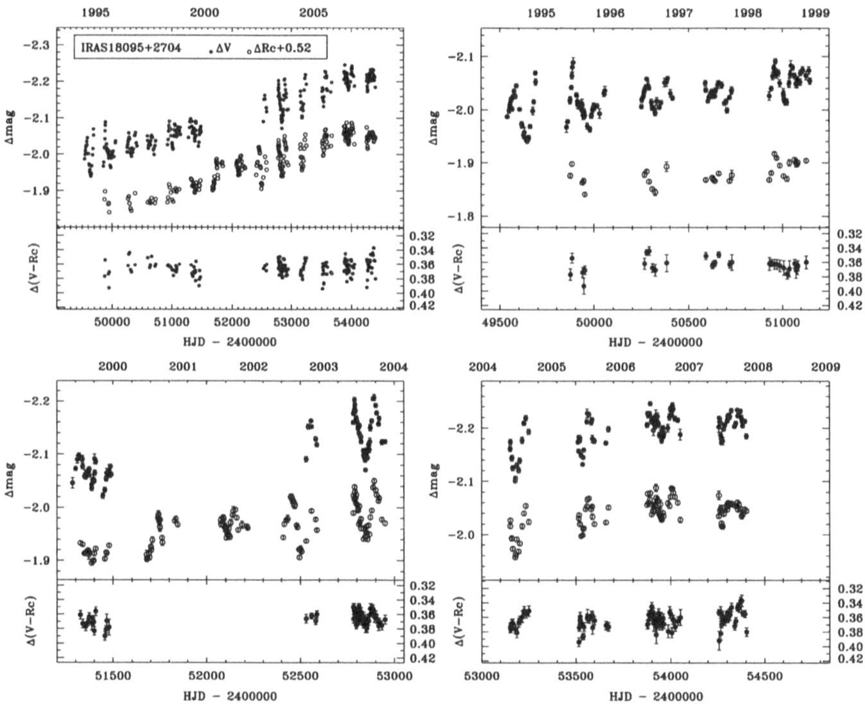

Fig. 2 *VR* light and color curves of the O-rich PPN IRAS 18095+2704 from 1994 to 2007. The *upper left panel* shows the entire light curve and the other panels show it in 5-year intervals. Note the cyclical variability with varying amplitude superimposed on a general increase in brightness

4 Extending the Studies

- *The Magellanic Clouds:* The *Spitzer Space Telescope* has greatly facilitated the identification of post-AGB stars in the Magellanic Clouds (MCs) based on their infrared dust spectra. We are using the OGLE dataset to investigate the variability in a sample of 17 C-rich PPNe and post-AGB objects [7]. Thus far, clear periodicity is found in two of them, with periods of 96 days (T = 6,250 K) and 74 days (F5-8 I) and amplitudes of 0.21 and 0.08 mag (*I*), respectively. Most of the others show short-timescale variations suggesting that they are hotter.
- *Oxygen-rich PPNe:* We have carried out observations of four O-rich PPNe since 1994. They are all F stars but show a range in periods from 43 to 114 days. They all have low amplitudes in the range 0.14–0.21 mags (*V*) and are each redder when fainter. This is illustrated in Fig. 1. They show a steeper rate of decrease of period with temperature than do the C-rich PPNe, as shown in Fig. 1, a property to investigate further when a larger sample is obtained.
- *Additional data for PPNe:* We have continued observing the initial 12 PPNe along with the O-rich ones mentioned above at the VUO. For several of these

there are also published data by Arkhipova and collaborators and for a few that are not too far north there exist data from the All Sky Automated Survey (ASAS). These additional data sets increase the density of the data and are especially helpful in the search for multiple-periods of pulsation in these objects. We have also begun observations of southern targets using the SARA telescopes.
- *Contemporaneous light and velocity observations:* Combining our light and color curves with radial velocity curves (Fig. 2) will allow us to study the pulsation in greater physical detail. We are carrying out an observational study of the radial velocity of these [5]. For one particular PPNe, IRAS 22272+5435, we have combined an extensive, published radial velocity study [8] with our data for the 2006–2007 seasons to produce well-defined light, color, and radial velocity curves. The object is observed to be brightest when hottest and smallest.

5 Goals and Implications

- *Physical Properties:* We are in the process of analyzing the contemporaneous light, color, and velocity using the Baade–Wesselink method to determine the radius of the star, and with this and other data to find its luminosity and mass. These will be the first direct determinations of these quantities for any PPNe; none have parallaxes or binary orbits.
- *Seeing Evolution in Real Time?* PPN are expected to evolve from the late-AGB to the early-PN phase in a time scale of a few thousand years [2]. Combining the rate of period change measured for the sample in PPNe in Fig. 1 with a post-AGB evolutionary model predicts a rate of change of −0.2 day/year. Investigation of data for the five PPNe with the best determined periods shows evidence of period changes but no systematic trends. This is complicated by the presence of multiple periods in the data. An additional decade of observations will hopefully help to determine or constrain the values of the evolutionary rate for these PPNe.
- *Multiple Periods:* The light curves often appear to show beat periods suggesting additional periods similar to the main period. Investigating our best-dermined cases reveals a period ratio (P_2/P_1) ranging from 0.85 to 1.02, with an average value of 0.95 based on six objects. This is significantly different from the value of 0.70 found for Cepheids. This should provide clues to the modes of pulsation.
- *Pulsational Models:* A glaring need exists for pulsational models for PPNe. Only a very limited set have been published [3] and new models are needed to compare with these new, detailed variability studies.

Acknowledgements We acknowledge the efforts of the many VU undergraduates who have carried out the observations to produce these light curves. This long-term research has been supported by the National Science Foundation (9315107, 9900846, 0407087, 1009974) and the Indiana Space Grant Consortium.

References

1. Arkhipova, V.P., Ikonnikova, N.P., Noskova, R.I., Sokol, G.V.: Astro. Lett. **26**, 609 (2000)
2. Blöcker, T.: A&A **297**, 755 (1995)
3. Fokin, A.B., Lèbre, A., Le Coroller, H., Gillet, D.: A&A **378**, 546 (2001)
4. Hrivnak, B.J., Lu, W., Maupin, R.E., Spitzbart, B.D.: ApJ **709**, 1042 (2011)
5. Hrivnak, B.J., Lu, W., Wefel, K.L., et al.: ApJ **734**, 25 (2011)
6. Kwok, S.: ARA&A **31**, 63 (1993)
7. Volk, K.M., Hrivnak, B.J., Matsurra, M., et al.: ApJ **735**, 127 (2011)
8. Zacs, L., Sperauskas, J., Musaev, F.A., et al.: ApJ **695**, 203 (2009)

A Testimator Approach to Detecting Heteroskedasticity in the Cepheid PL Relation

S.M. Kanbur, A. Nanthakumar, and C.C. Ngeow

Abstract The Cepheid Period–Luminosity (PL) relation is crucial to establish the size scale of the Universe independent of the Cosmic Microwave Background. A greater understanding of its systematics and underlying physics is therefore vital. Here we adapt the testimator approach, previously used to study a possible nonlinearity in the slope of the PL relation, to look at possible variations in the dispersion of the PL relation as a function of period—or heteroskedasticity. We describe the mathematics behind our new test and some test calculations. We then apply it to Large Magellanic Cloud Cepheid PL data using published reddenings, and show that there is a reduction in the variance of the PL relation around 10 days. While this could be due to errors in reddenings, analysis of multiphase PL relations suggest that this is a real feature. We close with a discussion of the possible implications of this result for the extra-galactic distance scale and stellar pulsation.

1 Introduction

The Cepheid Period–Luminosity (PL) relation is crucial to establish the size scale of the Universe independent of the Cosmic Microwave Background (CMB). A greater understanding of its systematics and underlying physics is therefore vital. In previous work, [2] refined the two-stage testimator procedure to multiple stages

S.M. Kanbur (✉)
Department of Physics, SUNY Oswego, Oswego, NY, 13126, USA
e-mail: shashi.kanbur@oswego.edu

A. Nanthakumar
Department of Mathematics, SUNY Oswego, Oswego, NY 13126, USA
e-mail: nanthaku@oswego.edu

C.C. Ngeow
Graduate Institute of Astronomy, National Central University, Jhongli, Taiwan
e-mail: cngeow@astro.ncu.edu.tw

Kanbur, S.M. et al.: *A Testimator Approach to Detecting Heteroskedasticity in the Cepheid PL Relation*. Astrophys Space Sci Proc. **31**, 127–130 (2013)
DOI 10.1007/978-3-642-29630-7_25,

to test for slope changes in the LMC PL relation. Using OGLE III data and published reddenings, they found evidence of a slope change around a period $\log P \sim 1.0$ for the mean light Cepheid PL relation.

In this paper, we adapt the testimator procedure to check for error variability (i.e. heteroskedasticity as opposed to homoskedasticity) in the context of simple regression and apply this test to the Cepheid PL relation at mean light and other phases. Full details of the procedure are given in [1]. Here we summarize the main ideas behind our statistical test.

2 The Testimator Approach

We assume data of the form $(X_i, Y_i), i = 1,,,n$, a simple linear regression model $Y = \alpha + \beta X + e$, where α is the y-intercept and β is the slope, and e is the random measurement error that is independent and normally distributed with mean 0 and variance σ_e^2. We note that the estimated slope, $\hat{\beta}$, and estimated error variance, $\hat{\sigma}_e^2$ are given by

$$\hat{\beta} = \frac{\sum_1^n (x_i - \bar{x})(y_i - \bar{y})}{\sum_1^n (x_i - \bar{x})^2}, \tag{1}$$

$$\hat{\sigma}_e^2 = \frac{1}{n-2} \sum_1^n (y_i - \hat{\alpha} - \hat{\beta} x_i)^2. \tag{2}$$

The basic idea is as follows: we arrange the data in order of increasing X_i and then divide the data into N bins. We perform a linear regression of the data in bin 1 and for the data in this first bin, estimate the variance using Eq. (2). Our null hypothesis is that for this bin, $H_0 : \sigma_e^2 = \sigma_0^2$ versus $H_A : \sigma_e^2 \neq \sigma_0^2$, where σ_0^2 is an initial estimate of the variance and the test-statistic is

$$\chi^2_{obs} = (n-2)\frac{\hat{\sigma}_e^2}{\sigma_0^2}. \tag{3}$$

Here, we define critical values

$$k_1^{(1)} = \frac{\chi^2_{obs}}{\chi^2_{\nu,\alpha/2m}}, k_2^{(1)} = \frac{\chi^2_{obs}}{\chi^2_{\nu,1-(\alpha/2m)}},$$

with $\nu = n - 2$ degrees of freedom and m is the number of bins the data is divided into. If $k_1^{(1)} > 1$ and $k_2^{(1})} < 1$ then reject H_0. If we do not reject H_0, define a smoothing constant as

$$k^{(1)} = \frac{k_1^{(1)}}{k_2^{(1)}},$$

and revise our estimate of σ_0^2 as follows

$$\sigma_{w1}^2 = k^{(1)}\hat{\sigma}_\epsilon^2 + (1 - k^{(1)})\sigma_0^2. \tag{4}$$

In a multi-stage process, we repeat this procedure many times. Whenever H_0 is rejected, we have evidence of a change of error variance, or heteroskedasticity, and moreover this change occurs in the range of independent variable covered by the bin at which H_0 was rejected. In the second stage, we use the data from the second bin to test the hypothesis $H_0 : \sigma_e^2 = \sigma_{wl}^2$ verses $H_A : \sigma_e^2 \neq \sigma_{wl}^2$. We construct a linear regression using the data in bin 2 and formulate an estimate of the error variance using Eq. (2) and the second data set. A second test statistic is formulated using Eq. (3) with corresponding critical points $k_1^{(2)}, k_2^{(2)}$ as per the first stage. Define $k^{(2)}$ similarly to the first stage and reject the null hypothesis if $k_1^{(2)} > 1$ and $k_2^{(2)} < 1$. If this is not true, then accept H_0 and move onto the next stage by first formulating another "smoothed" estimate of the error variance using Eq. (3).

The testimator approach described above has a number of advantages. Firstly, under H_0, the testimator is unbiased about the error variance. Secondly, by using the Bonferroni type testing procedure, the probability of a family wide error is less than or equal to α. Thirdly, at every stage in the multi-step process, the variance of the testimator estimate of the variance is strictly less than the variance of the standard estimator of variance.

3 Results

Initially we create two sets of simulated data $(X_i, Y_i), i = 1,,, N$, one with constant error variance and another with varying error variance within a small range of the independent variable. The multi-stage testimator process correctly finds the heteroskedastic dataset and the range of the independent variable where this varying error variance is found.

The we use OGLE III V and I band LMC PL data at mean light corrected for reddening as [3]. The testimator procedure finds evidence of heteroskedasticity in the V band relation and marginal results in the I band.

4 Multiphase Results

The results in the previous section are for mean light PL relations—that is the relationship between period and the average magnitude over a pulsation cycle. We can also plot multiphase PL relations by re-phasing all light curves to a common starting point. Plotting Cepheid PL relations as a function of phase demonstrates clearly the varying and dynamic nature of the PL relation. Because the reddening is the same at all these phases, any observed changes must be intrinsic. The mean light PL relation is an average over the relations at these phases. Of interest here is the fact that the variance of the PL relation changes rapidly both with pulsation phase and with period.

5 Conclusion

The possible heteroskedasticity of the Cepheid PL relation may have a small effect on the extra-galactic distance scale. However, it does present a major challenge to theories of stellar pulsation. Our multiphase results suggest that this changing variance is real and not caused by errors in reddening.

Acknowledgements SMK and SR thank SUNY Oswego for supporting this research. CCN thanks the funding from National Science Council (of Taiwan) under contract NSC 98-2112-M-008-013-MY3.

References

1. Kanbur, S., Nanthakumar, A., Nanthakumar, C., Ngeow, A. Marsh. (in press)
2. Kanbur, S., Ngeow, C., Nanthakumar, A., Stevens, R.: PASP **119**, 512–520 (2007)
3. Ngeow, C., Kanbur, S.M., et al.: AJ **693**, 691 (2011)

Synthetic Cepheid Period–Luminosity and Period–Color Relations in *Spitzer's* IRAC Bands

C.-C. Ngeow, M. Marconi, I. Musella, M. Cignoni, M. Marengo, and S.M. Kanbur

Abstract The mid-infrared period–luminosity (P–L) relation for Cepheids will be important in the *JWST* era, as it holds the promise of deriving the Hubble constant within 2% accuracy. We present the synthetic P–L and period–color (P–C) relations, derived from a series of stellar pulsation models with varying helium and metallicity abundance, in the *Spitzer* IRAC bands. Selected synthetic P–L and P–C relations were compared to the empirical relations derived from Galactic and Large Magellanic Cloud Cepheids. We also present the empirical P–L relations based on the latest distance measurements using infrared surface brightness techniques.

1 Introduction

The period–luminosity (P–L, also known as Leavitt Law) relation for classical Cepheid variables (hereafter Cepheids) is an important tool in astrophysics, as it is

C.-C. Ngeow (✉)
Graduate Institute of Astronomy, National Central University, Jhongli City 32001, Taiwan
e-mail: cngeow@astro.ncu.edu.tw

M. Marconi · I. Musella
Osservatorio Astronomico di Capodimonte, Via Moiariello 16, 80131 Napoli, Italy
e-mail: marcella.marconi@oacn.inaf.it; ilaria@na.astro.it

M. Cignoni
Department of Astronomy, Bologna University, via Ranzani 1, 40127 Bologna, Italy
e-mail: michele.cignoni@unibo.it

M. Marengo
Department of Physics, Iowa State University, Ames, IA 50011, USA
e-mail: mmarengo@iastate.edu

S.M. Kanbur
Department of Physics, State University of New York at Oswego, Oswego, NY 13126, USA
e-mail: smkanbur@gmail.com

Ngeow, C.-C. et al.: *Synthetic Cepheid Period–Luminosity and Period–Color Relations in* Spitzer's *IRAC Bands*. Astrophys Space Sci Proc. **31**, 131–135 (2013)
DOI 10.1007/978-3-642-29630-7_26,

the first rung in distance scale ladder, and can be used to constraint stellar pulsation models. Investigations of Cepheid P–L relations, both empirically and theoretically, have been conducted mainly in optical and near infrared bands (for example, see [1] and reference therein). Recently, researches on Cepheid P–L relations have been expanded to the mid-infrared (MIR), as MIR P–L relations will be important in the *James Webb Space Telescope (JWST)* era. Observations with *Spitzer* and/or *JWST* in MIR offer reductions of several key systematic errors in Hubble constant.[1] This opens the possibility of deriving the Hubble constant, via the cosmic distance ladder, to $\sim$ 2% level [2, 3]. Several empirical studies of MIR P–L relations in *Spitzer's* IRAC bands have already been presented in [4–10]. In terms of theoretical investigation, [11] present the synthetic IRAC band P–L and period–color (P–C) relations constructed from state-of-the-art stellar pulsation codes. The aim of this paper is to provide an overview of these synthetic P–L and P–C relations, and compare them to the empirical P–L and P–C relations.

2 The Synthetic P–L and P–C Relations

The synthetic P–L and P–C relations were generated from a series of stellar pulsation models using different inputs of metallicity (Z) and helium abundance (Y). Procedures and input physics for these stellar pulsation models are same as those used for theoretical investigation of optical and near infrared P–L relations, as presented in [1], with the exception that bolometric light curves were converted to IRAC band magnitudes using model atmospheres. In brief, the non-linear stellar pulsation models used in this work include a treatment of time-dependent coupling between pulsation and convection, and a canonical mass-luminosity relation. About 10^3 pulsators generated from the pulsation models were used to populate the instability strip according to a mass law. Further details can be found in [11] and reference therein. Figure 1 presents an example of the synthetic P–L and P–C relations from one of the model sets.

3 Comparison to Observations

Marengo et al. [8] presented single-epoch *Spitzer* observations for about 29 nearby Galactic Cepheids, and derived the empirical P–L relations for these Cepheids using three sets of distance measurements. These include the Baade–Wesselink type infrared surface brightness (IRSB) techniques with either an "old" or a "new" projection (p) factor, and geometrical parallaxes measurements. In addition, [12] has also derived the IRAC band P–L relations for these Cepheids based on distance measurements using a calibrated Wesenheit function. Finally, the latest IRSB distances for these Cepheids are available from [13], with a steeper p-factor relation. Hence, we re-derive the P–L relations using either the geometric parallaxes, if

[1]For example, one of systematic errors being the extinction, as it is negligible in MIR.

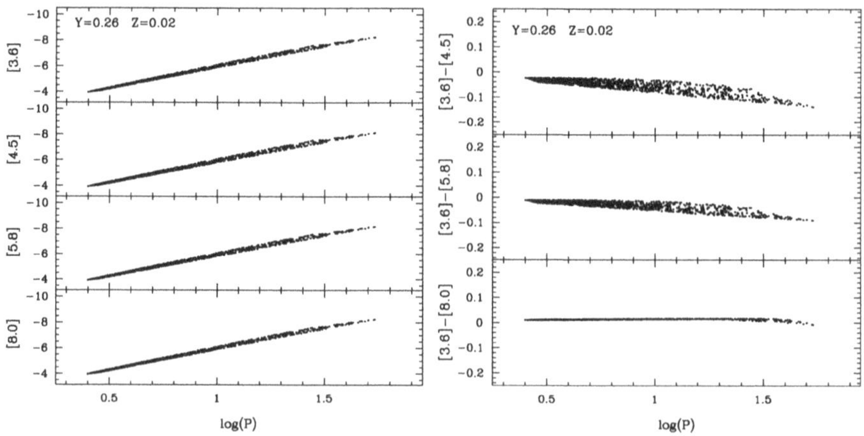

Fig. 1 Example of synthetic P–L relations (*left panels*) and P–C relations (*right panels*) based on $Y = 0.26$, $Z = 0.02$ model set

Table 1 New empirical IRAC band P–L relations for Galactic and LMC Cepheids

Band:	3.6 μm	4.5 μm	5.8 μm	8.0 μm
		Galactic Cepheid		
Slopes	−3.303 ± 0.031	−3.216 ± 0.031	−3.257 ± 0.027	−3.299 ± 0.025
Intercepts	−2.429 ± 0.034	−2.482 ± 0.035	−2.414 ± 0.030	−2.411 ± 0.028
		LMC Cepheid		
Slopes	−3.268 ± 0.042	−3.142 ± 0.042	...	...
Intercepts	−2.544 ± 0.061	−2.682 ± 0.061	...	...

available, or these latest IRSB distances. The derived P–L relations are summarized in Table 1. These five sets of empirical P–L slopes are compared to the selected synthetic P–L slopes in left panel of Fig. 2. It is worth to point out that the empirical P–L slopes become shallower and closer to those based on geometrical parallaxes when the adopted p-factor relation is steeper in IRSB techniques. Right panel of Fig. 2 compares the synthetic P–C relations with empirical result, disagreement can be seen between synthetic and empirical P–C relations.

Empirical P–L relations for Large Magellanic Cloud (LMC) Cepheids were compared to the synthetic P–L slopes from selected models sets in left panel of Fig. 3. These empirical P–L slopes include those presented in [7] and [6, 10]. Note that the 3.6 μm & 4.5 μm band mean magnitudes given in [10] are based on the available full light curves of 67 LMC Cepheids, and 29 of them have new IRSB distance measurements from [14]. The derived P–L relations by combining [10] photometry and [14] distances, for Cepheids with period between 10 and 60 days, are summarized in Table 1 and compared to the synthetic P–L slopes in Fig. 3. Right panel of Fig. 3 compares the selected synthetic P–C relations to the empirical P–C relation given in [10]. See [11] for further discussion of the comparison of these P–C relations.

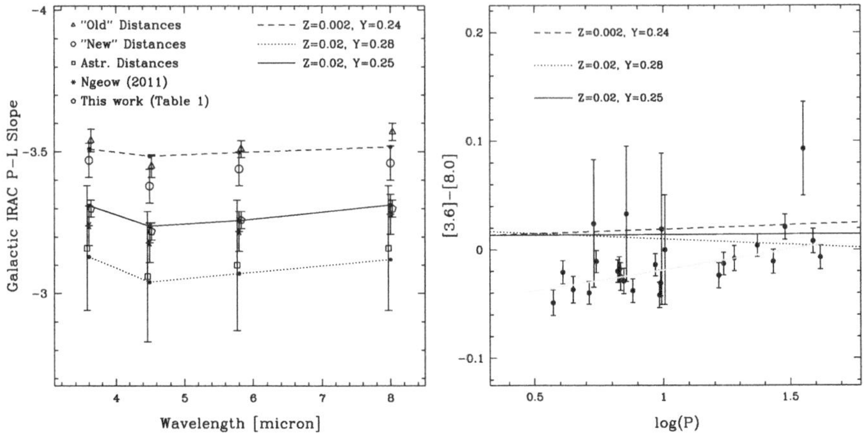

Fig. 2 *Left Panel:* Comparison of the empirical Galactic P–L slopes to the synthetic P–L slopes from selected model sets. The P–L slopes based on "Old", "New" and Astr distances are adopted from [8]. Wavelengths of the data points have been shifted slightly for better visualization. *Right Panel:* Comparison of the synthetic P–C relations to the empirical Galactic P–C relation, where the data points are taken from [8]. The *cyan* (or *gray*) *line* is the fitted P–C relation to the data [8]

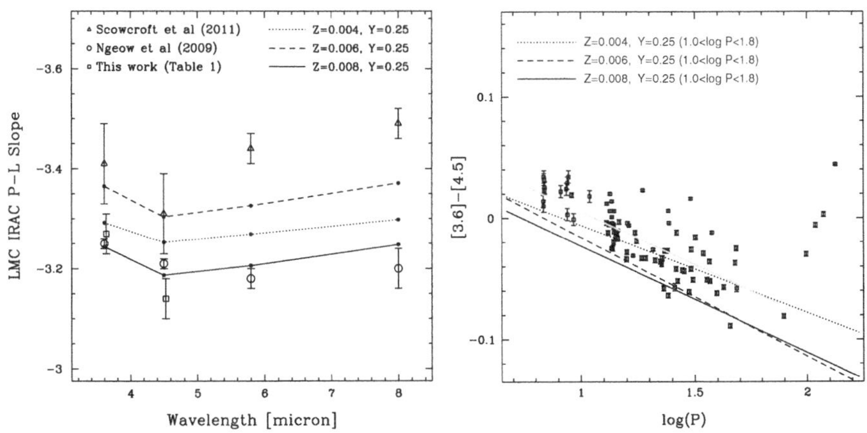

Fig. 3 *Left Panel:* Comparison of the empirical LMC P–L slopes to the synthetic P–L slopes from selected model sets. Scowcroft et al. [10] P–L slopes include those from [10] (for 3.6 μm & 4.5 μm band) and [6] (for 5.8 μm & 8.0 μm band). Wavelengths of the data points have been shifted slightly for better visualization. *Right Panel:* Comparison of the synthetic P–C relations to the empirical LMC P–C relation, where the data points are taken from [10]. The *cyan* (or *gray*) *line* is the fitted P–C relation to the data ([10])

Acknowledgements CCN thanks the funding from National Science Council (of Taiwan) under the contract NSC 98-2112-M-008-013-MY3.

References

1. Bono, G., Caputo, F., Marconi, M., Musella, I.: ApJ **715**, 277 (2010)
2. Freedman, W.L., Madore, B.F.: Ann. Rev. Astron. Astrophys. **48**, 673 (2010)
3. Freedman, W.L., Madore, B.F., Scowcroft, V., et al.: (2011) arXiv:1109.3802
4. Freedman, W.L., Madore, B.F., Rigby, J., Persson, S.E., Sturch, L.: ApJ **679**, 71 (2008)
5. Ngeow, C., Kanbur, S.M.: ApJ **679**, 76 (2008)
6. Madore, B.F., Freedman, W.L., Rigby, J., et al.: ApJ **695**, 988 (2009)
7. Ngeow, C.-C., Kanbur, S.M., Neilson, H.R., et al.: ApJ **693**, 691 (2009)
8. Marengo, M., Evans, N.R., Barmby, P., et al.: ApJ **709**, 120 (2010)
9. Ngeow, C.-C., Kanbur, S.M.: ApJ **720**, 626 (2010)
10. Scowcroft, V., Freedman, W., Madore, B.F., et al.: (2011) arXiv:1108.4672
11. Ngeow, C.-C., Marconi, M., Musella, I., et al.: (2011) arXiv: 1111.2100N
12. Ngeow, C.-C.: ApJ **747**, 50 (2012)
13. Storm, J., Gieren, W., Fouqué, P., et al.: A&A **534**, A94 (2011)
14. Storm, J., Gieren, W., Fouqué, P., et al.: A&A **534**, A95 (2011)

Part III
New Instrumentation from the Ground and Space

New Ground-Based Observational Methods and Instrumentation for Asteroseismology

Pedro J. Amado

Abstract Space instrumentation like SOHO, MOST, CoRoT & Kepler has been and is being built to attain very high precision data to be used for asteroseismic analysis. Nonetheless, there is a very strong need for providing additional information, especially on mode identification. With this contribution I will review the efforts been put on new ground-based instrumentation and the methodology that can be used to achieve this aim.

1 Introduction

Topics like rotation, convection, diffusion and settling of heavy elements vs. mixing, magnetic fields and others are a challenge for both stellar structure and evolution and for asteroseismology. These issues, not being fully understood, limit our ability to determine accurate stellar ages, global parameters and the details of the internal structure of the stars.

To help solving the aforementioned challenges, asteroseismology needs the observational input from ground-based instrumentation. However, for asteroseismology to work, the observational inputs need to be as accurate as possible, so that a comparison and refinement of models can be achieved. This observational input can be classified in three groups: (1) *Physical parameters of pulsating stars*, for which ground-based instrumentation has provided a wealth of information for all type of stars in the HR diagram; (2) *pulsation frequencies*, whose quality depends on the ability of the data time series to resolve all the frequencies and determine them with enough precision; and (3) *mode identification*, which requires the comparison between the oscillation theory applied to the stellar atmosphere and

P.J. Amado (✉)
Instituto de Astrofísica de Andalucía, Glorieta de la Astronomía s/n, Granada, Spain
e-mail: pja@iaa.es

Amado, P.J.: *New Ground-Based Observational Methods and Instrumentation for Asteroseismology*. Astrophys Space Sci Proc. **31**, 139–144 (2013)
DOI 10.1007/978-3-642-29630-7_27,

the asteroseismic observables like frequencies, their amplitudes and phases or line profile variations (LPVs). Mode identification is probably the most important input asteroseismology needs.

2 Mode Identification

The approaches followed to tackle this problem can be grouped into three different categories: (1) *Frequency patterns:* in the asymptotic regime, the large and small separation in solar-like oscillations or the use of the Frequency Ratio Method [11] for γ Doradus stars help reducing the values of the degree and radial order of the modes (see Reed et al., these proceedings). Also, for δ Scuti stars, quasi-periodicities are been found in a handful of objects which are predicted by the theory, but these periodicities do not necessarily provide mode identification, although it can be used as an additional asteroseismic observable (see García-Hernández et al. and Suárez et al., these proceedings); (2) *direct fitting* of observed and computed frequencies considering all quantum number values is difficult, especially when small or very large number of frequencies are observed, or when they do not show frequency patterns; and (3) *mode identification of a single or a very few modes*, especially of the main pulsating modes in the star, can imply a large reduction of models fitting the observations. This last approach is the one that can benefit most from ground-based observations.

To carry out the identification of the single modes there are two types of data which need to be taken as time series: (1) *multicolour photometry*, which provides information from the different amplitude and phase behaviour of the modes at different wavelengths bands, thus providing information of the degree ℓ of the mode ([7]; see also Pamyatnykh et al. in these proceedings); and (2) *high-resolution spectroscopy* allows us to study LPVs with a set of different methods.

To study LPVs time series from these spectroscopic data, two approaches can be followed: (1) *to careful select a number of lines*, which depends on the spectral type of the star [5, 17], they must be unblended, deep, thermally broadened (avoid hydrogen lines), etc; and (2) *averaging all lines in the spectrum* [15] providing a high SNR mean line profile by using optimization techniques like the LSD [6], cross-correlation, etc. The drawback of this latter method is that it assumes that all the lines have the same intrinsic profile and that they form in the same region of the atmosphere, which may not be the case.

3 Past and Current Instrumentation for Asteroseismology

The various types of data and techniques mentioned above have been used for a long time. For instance, the discovery of periodic variable stars begun back in the seventeenth century and Struve was already detecting LPVs in the spectrum of the star δ Sct in 1953 [14]. The list of instruments providing these data is enormous and, therefore, just a few milestones that have occurred in the past few years will be briefly mention: (1) *Small and medium-sized telescopes* all over the world have been collecting optical photometry or spectroscopy, including APTs, in a variety of photometric indices (Johnson, Strömgren, Geneva, etc) with photoelectric photometers and with CCDs detectors; (2) *coordinated networks* set up by both the Solar (GONG, BiSON, IRIS, TON) and the Stellar (WET, DSN, STEPHI or multi-site campaigns with different telescopes) communities have been used to obtain observations all around the world. They have improved the precision in the determination of the frequencies, avoiding strong aliases problems with much cleaner spectral windows; (3) *development of high-accuracy RV spectrographs*, which have allowed in the past few years the detection and eventually the confirmation of the presence of solar-like oscillations in stars other than the Sun: HARPS, UVES, CORALIE, SARG, UCLES, SOPHIE, FIES, HERMES, HERCULES, etc. They have contributed to the refinement of the stars' global parameters and provided mode identification. However, there have been very few coordinated multi-site campaigns in the last years, two of which are the spectroscopic follow up for CoRoT ([16]; Poretti et al., these proceedings) or the coordinated campaign for Procyon (11 spectrographs; [3]); (4) *Large Surveys*: OGLE, MACHO, RATS, irsf/sirius, VVV,[1] etc; finally (5) *Interferometry* is a very young field with a bright future which can contribute to asteroseismic studies by providing accurate radii for stars and perhaps much more (see contributions in these proceedings on VLTI, SUSI, KI, PTI, NPOI, CHARA, COAST).[2]

All these instruments and projects have contributed to the three basic inputs needed for asteroseismology named in Sect. 2. However, there are two main areas that have not been fully exploited by current instrumentation for asteroseismology, viz., stellar networks for high-resolution spectroscopy or from the single high-duty-cycle site of Antarctica, and the NIR wavelength observations (ultraviolet is another unexplored wavelength region but it needs space instrumentation).

Long, high-duty-cycle RV/LPVs time series: The solution for obtaining these high-quality time series is either to implement a worldwide network of hires, high-accuracy spectrographs or to implement a single one in a site like Antarctica. An example of the first is SONG (Stellar Observations Network Group; [8]) which will set up a network of eight alt-azimuthal 1-m telescopes providing 1 m/s precision radial velocities, and photometry of crowded fields with a dual-channel lucky imaging camera with up to six filters. An example of the second is SIAMOIS (Seismic Interferometer Aiming to Measure Oscillations in the Interior of Stars; [10]), which is a project to achieve high precision Doppler velocities from Dome C that could carry out a scientific programme complementary to CoRoT and

---

[1] http://mwm.astro.puc.cl/mw/index.php/Main_Page

[2] http://en.wikipedia.org/wiki/List_of_astronomical_interferometers_at_visible_and_infrared_wavelengths

Kepler. The core of the instrument, providing the necessary sensitivity and stability, is a Fourier tachometer similar to the helioseismic network GONG, fully automated and without moving parts. The high duty cycle (up to 90%) is comparable to space-borne observations.

Near-Infrared wavelength observations for asteroseismology: A list of reasons for the interest of observing pulsating stars in the near infrared (NIR) is: (1) it will provide a new wavelength region from which to obtain stellar physical parameters, especially for red giants, which show a sufficient number of lines and molecular bands; (2) the flux peak moves toward NIR wavelengths for cooler objects, providing thus higher SNR data; (3) red giants show larger amplitudes in the NIR than at VIS wavelengths; (4) NIR observations probe different parts of the line forming region in stellar atmospheres with respect to the VIS wavelengths. Therefore, they can improve mode identification.

The main reason to explain why there have been no systematic observations of pulsating stars in the NIR wavelength regions is because of the poor precision achieved with NIR detectors. This is no longer the case, as absolute photometric accuracies better than 10 mmag can be achieved in the JHK bands [9, 13]. Other reasons for using NIR photometry are: (1) NIR commercial detectors allow two-channel cameras to simultaneously observe in the VIS and in the NIR in small telescopes; (2) red and NIR wavelengths provide increased discrimination of pulsating modes [4]; (3) the Y band is accessible without cryogenics, JH bands need cooling though. Reasons for using NIR spectroscopy are: (1) there is less crowding of lines, implying less blending in high-velocity rotators which in turns means that LPVs are more clearly detected; (2) NIR observations are expected to provide increased discrimination of pulsating modes in spectroscopy through differences in amplitudes and phases in the NIR when compared with the VIS [1,2].

To exploit the increased mode discrimination of red and NIR wavelengths, photometric instruments should provide simultaneous VIS and NIR light curves. A few examples of current and future instrumentation are: REM@LaSilla is a 60 cm diameter fast reacting telescope with two instruments REMIR, an infrared imaging camera, and ROSS, a visible imager and slitless spectrograph which can work simultaneously. GROND at MPI/ESO2.2m@LaSilla is an imaging instrument to take observations simultaneously in four visible and three NIR channels. Bootes IR@OSN in Granada is another 60 cm diameter fast reacting telescope with a CCD detector and soon will be equipped with an IR camera working in the NIR bands. MAIA@MERCATOR will provide fast photometry in three simultaneous wavebands in the SDSS system u', g' r'+i'. PANIC@2.2m of CAHA is a panoramic camera in the NIR with a mosaic of four Hawaii-2RG IR detectors which will provide accurate NIR light curves of variable stars in a very wide FoV.

However, the azimuthal order, m, cannot be determined by photometric means alone. Each mode of pulsation introduces a distinctive velocity field on the surface of the star. This is observed in the stellar spectrum as a periodic variation in the line profiles with high resolution spectroscopy. And it is here where NIR spectroscopy can make a real change. However, the existing high resolution NIR spectrographs do not have cross-dispersion, covering therefore a very small range in wavelength, or do not have sufficient resolution for RV and LPV work. Some future projects will deliver instruments providing high resolution and large wavelength coverage. CARMENES (Calar Alto high-Resolution search for M dwarfs with Exo-earths with Near-infrared and optical Echelle Spectrographs; [12]) is one of these instruments. It is a next-generation instrument to be built for the 3.5 m telescope at the Calar Alto Observatory by a consortium of eleven Spanish and German institutions. Conducting a 5-year exoplanet survey targeting 300 M dwarfs with the completed instrument is an integral part of the project. The CARMENES instrument consists of two separate echelle spectrographs covering the wavelength range from 550 to 1,700 nm at a spectral resolution of R = 82,000, fed by fibers from the Cassegrain focus of the telescope. The spectrographs are housed in vacuum tanks providing the temperature-stabilized environments necessary to enable a 1 m/s radial velocity precision employing a simultaneous calibration with an emission-line lamp. For late-M spectral types, the wavelength range around 1,000 nm (*Y* band) is the most important wavelength region for radial velocity work. The instrument concept is based in a design with two spectrographs, one equipped with a CCD for the range 550–1,050 nm, and one with HgCdTe detectors for the range from 900 to 1,700 nm.

CARMENES asteroseismology science cases are: (1) mode identification in the NIR, (2) Pulsations in M dwarfs (see Rodríguez-López et al. and Moya et al., these contributions), (3) Pulsating stars with planets (see Moya et al. and M. Oshagh et al., these proceedings), (4) Cepheids (Nardetto et al., these proceedings), (5) Solar-like oscillations in red giants (De Ridder et al., Hekker et al., these proceedings).

4 Future of Ground-Based Instruments Useful for Asteroseismology: Conclusions and Final Remarks

Unfortunately, there is not much to list as future ground-based instrumentation for asteroseismology due to the fact that, either this author is missing part of the instrumental development for asteroseismology for the near future, or there is not much effort/resources being put into this area. Synergies with exoplanet research is still providing very interesting projects but probably little time will be awarded to pure asteroseismic research. These projects involve the development of visible spectrographs like ESPRESSO@VLT, CODEX@E-ELT and NIR spectrographs like CARMENES, SpiRou and UPF. Interferometric instrument is another area of development useful for asteroseismology research.

Ground-based instrumentation is essential for asteroseismology. The determination of accurate physical parameters, especially for science preparation of expensive space missions, need more resources for a homogeneous determination and for cross-calibrations with different methods: photometry, spectroscopy, interferometry, etc. Space photometry provides accurate pulsation frequencies. However, solar-like oscillations can benefit from VIS and NIR hires spectroscopy, especially for red giants. There are new areas of development still to be exploited: spectroscopic networks in the VIS, and NIR spectrographs which can lead to the development of VIS+NIR spectroscopic networks.

Space missions like CoRoT, Kepler and maybe PLATO are and will continue to revolutionize the field of asteroseismology. Nonetheless, a few final questions put forward by the author for discussions are the following: are the resources invested in space instrumentation followed by sufficient resources for ground-based instrumentation? And for new theoretical developments (human resources)? What would be the results for asteroseismology if similar resources to those allocated for space missions were used in the aforementioned items? Could this ground-based instrumentation exploit methods which have not yet been fully tested? Is the access of the asteroseismic community to very large telescope sufficient? Exoplanet research has driven asteroseismology research through the synergies in techniques and instrumentation, but how much can we exploit those synergies? and, is asteroseismology research being held back by the limitations impose by exoplanet research? Is the asteroseismic community strong enough to drive the development of their own specific instrumental project roadmap? Answer to these questions might be provided in the near future.

Acknowledgements The author acknowledges financial support from projects AYA2009-08481-E and AYA2010-14840 from the Spanish Ministry of Science and Innovation.

References

1. Amado, P.J., Suárez, J.C., Garrido, R., et al.: CoAst **150**, 311 (2007)
2. Amado, P.J.: CoAst **151**, 57 (2007)
3. Arentoft, T., Kjeldsen, H., Bedding, T.R., et al.: ApJ **687**, 1180 (2008)
4. Balona, L.A, Evers, E.A.: MNRAS **302**, 349 (1999)
5. De Ridder, J., Dupret, M.-A., Neuforge, C., et al.: A&A, **385**, 572 (2002)
6. Donati,J.-F., Semel,M., Carter,B.D., et al.: MNRAS **291**, 658 (1997)
7. Garrido, R., Garcia-Lobo, E., Rodriguez, E.: A&A **234**, 262 (1990)
8. Grundahl, F., Kjeldsen, H., Frandsen, S., et al.: MmSAI **77**, 458 (2006)
9. Minniti, D., Lucas, P.W., Emerson, J.P., et al.: NewA **15**, 433 (2010)
10. Mosser, B., Buey, T., Catala, C., et al.: CoAst **158**, 337 (2009)
11. Moya, A., Suárez, J.C., Amado, P.J., et al.: A&A **432**, 189 (2005)
12. Quirrenbach, A., Amado, P.J., and the CARMENES consortium: SPIE **7735E**, 37 (2010)
13. Sollima, A., Cacciari, C., Arkharov, A.A.H., et al.: MNRAS **384**, 1583 (2008)
14. Struve, O.: PASP **65**, 252 (1953)
15. Uytterhoeven, K., Mathias, P., Poretti, E., et al.: A&A **489**, 1213 (2008)
16. Uytterhoeven, K., Poretti, E., Mathias, P., et al.: AIPC **1170**, 327 (2009)
17. Zima, W.: A&A **455**, 227 (2006)

Impact of High Precision Photometry from Space: Do the Results Meet the Expectations?

Eric Michel

Abstract For decades already, our community is working at developing stellar seismology as a unique technique to reach a new insight into stars and address problems hitherto out of reach. In this framework, space photometry projects have been dedicated great efforts and have raised great expectations. In the light of recent results, I show here to which extent this approach is keeping its promises, confirming space photometry monitoring as a relevant approach in our present and future strategy.

1 What Were/Are These Expectations?

Prospective studies have long established how the seismic parameters characterizing eigenmodes (frequencies, amplitudes, lifetimes, phases,...) are sensitive to the very details of stellar interiors and to the possible effect of various phenomena, if measured with sufficient precision.

These new observables were shown very promising to test the "microphysics" involved in stars (opacities, equation of states, nuclear reaction rates,...), but also hydrodynamic processes hitherto neglected in standard models because of the lack of observational constraints: transport of chemical species, convective heat transport near the surface, rotation and its effect on structure and mixing,...

Simultaneously, great efforts have been spent at obtaining the data suited for seismic interpretation. Specific requirements in terms of signal-to-noise (S/N), but also sampling, duration, duty cycle, imply dedicated observational projects, generally in an international framework. Beside efforts put in ground-based networks, space soon appeared to be a promising place to obtain long, precise and uninterrupted photometry data. Since the 1980s, our community always had at least one space project in preparation [36].

E. Michel (✉)
Observatoire de Paris- LESIA,UMR8109, pl. J. Janssen 92195 Meudon, France
e-mail: Eric.Michel@obspm.fr

Michel, E.: *Impact of High Precision Photometry from Space: Do the Results Meet the Expectations?*. Astrophys Space Sci Proc. **31**, 145–150 (2013)
DOI 10.1007/978-3-642-29630-7_28,

Table 1 A few indicative numbers about the data made available by space photometry projects

Observed today	MOST	CoRoT	Kepler
Number of stars	∼480	∼130 +(∼130,000)	∼150,000
m_V range	$5.4 \leq m_V \leq 11$	$5.4 \leq m_V \leq 9.5$ ($11 \leq m_V \leq 16$)	$7 \leq m_V \leq 13$
Sampling time	∼10 s to 1 mn	1s (512 s, 32 s for ∼10%)	30 mn (1 mn for ∼4,000 stars)
$\sigma_{measurement}$	$\sigma_{1.5s} \sim 3\ 10^{-3}$ for a $m_V \sim 7$[a]	$4\ 10^{-4} < \sigma_{1s} < 2\ 10^{-3}$ ($3\ 10^{-4} < \sigma_{512s} < 1.5\ 10^{-3}$)	$1.5\ 10^{-5} < \sigma_{1mn} < 2.6\ 10^{-4}$
Duration of runs	∼10–60 days	40% long runs (140–170 days), 24% intermediate runs (60–80 days), 36% short runs (20–30 days)	Over 1.5 year already for most of the long cadence and for ∼180 short cadence targets, ∼ 1 month for the rest
Duty cycle	35%–80%	90%	90%

[a] From [35]

2 Unprecedented Light Curves

MOST (2003) was the first mission dedicated to seismology to fly and to confirm the space potential in terms of duty cycles. It has observed hundreds of stars already (including guide stars), over periods up to 2 months [25]. While it soon showed its limitation to track micromagnitude phenomena like solar-like pulsations [8], MOST has the interesting specificity to allow pointing almost everywhere in the sky and thus track rare specific targets, even if at the cost of a limited duty cycle (see e.g. roAp stars, [19]).

CoRoT (2006) and its polar inertial orbit made possible uninterrupted runs up to 6 months. Its 27 cm aperture and photon noise dominated measurements revealed tiny flux variations, including the searched solar-like oscillations in Main Sequence stars [28]. About 130 "bright" stars and 130,000 fainter ones have already been observed covering a large part of the HR diagram [3]. One specificity of CoRoT resides in the fact that its 4 deg^2 field of view (FOV) can be selected within disks of ∼10 deg radius, in two opposite directions [2].

Kepler, with its 95cm aperture and its 105 deg^2 FOV [21], has already observed ∼150,000 stars. One main specificity of the Kepler light curves is their length. Thanks to its Earth-trailing heliocentric orbit, light curves as long as 1.5 years are already available...

Table 1 gives a flavor of this impressive harvest with indicative numbers.

3 Unprecedented Mode Parameters

The motivation for such high S/N, high duty cycle, long duration light curves was to derive precise mode parameters to be used for comparison with models. In this respect, solar-like pulsators, because of their intrinsically low amplitudes and of the stochastic nature of the excitation have played a dimensioning role.

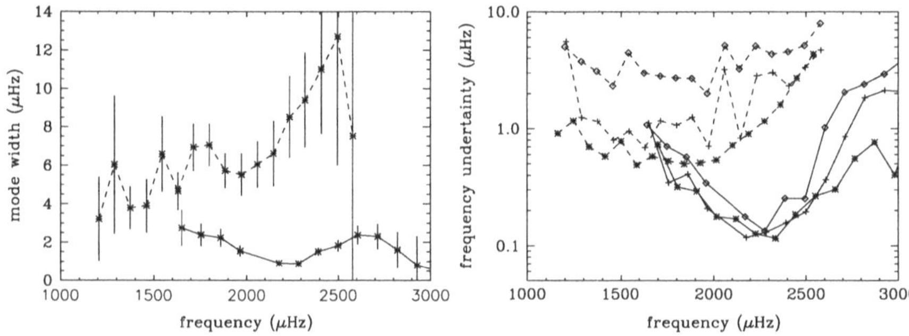

Fig. 1 *Left*: modes width (l=1) for HD 43587 (T_{eff} ~5,900 K, *solid line*) and HD 49933 (T_{eff} ~6,750 K, *dash line*); *right*: frequency uncertainties (l=0: *cross*; l=1: *stars*; l=2: *rhombs*) for HD 43587 (*solid line*) and HD 49933 (*dash line*). Results from [10] and preliminary study by the CoRoT DAT

Preliminary studies suggested that a precision of the order of 0.1 μHz on individual eigenfrequencies determination was the goal to pursue to have access to fine structural features of these objects (see [27] and references therein).

This goal has been achieved for early G Main Sequence solar-like pulsators (see Fig. 1), and for more evolved stars like subgiants or red giants [12, 16].

For early F Main Sequence solar-like pulsators, the error bars obtained so far are higher than anticipated (~1 μHz see Fig. 1) because of the unexpected large linewidths revealed by CoRoT for these hotter stars [1].

For opacity driven pulsators, where we deal with self-sustaining modes, as anticipated, conditions are more favorable and precise frequencies and amplitudes are now available for large sets of modes [34].

4 First Interpretations vs Scientific Objectives

While observations and analysis are still in progress, the phase of modeling and interpretation is developing rapidly. Though this phase is still young, it becomes more and more enlightening to see how these individual results take place in front of our scientific objectives. We select here a few illustrative cases and comment them briefly within the limited space available.

Extension of mixed cores: Characterizing the extension of mixed cores, beyond the formally convective region is a key objective because of the role of tank this region plays for the nuclear reaction material and its crucial impact on the evolution pace and stellar age. No quantitative theoretical description exists and we need observational constraints.

Several studies address this point already. They have been led on Main Sequence stars with various masses, from ~1.2 $M_\odot$ (F star HD 49933, [11]), to ~7$M_\odot$ (B star HD 50230, [15]), but also for more evolved stars, like the post-Main Sequence

$\sim$1.2M$_\odot$ G star HD 49385 [17]. Most of these studies point toward the need for a large extra mixing distance $d_{ov} \geq 0.2$ Hp (Hp the pressure scale height). Beyond this first indication, the most interesting is expected from the cross-comparison of these results and the ones to come, since the underlying processes are probably too complex to be handled in models with a single parameter description as they are now. In this perspective, the increasing number of stars bringing separate diagnostics for confrontation is very promising.

Structure of the outer convective zone: A better understanding and modeling of this region is necessary to secure the link between the observed surface properties and the inner part of the star. This point is addressed under various angles already.

Sharp variations of the sound speed are known to introduce oscillatory signature in the frequencies. Thanks to space photometry data, the use of this signature, long limited to the Sun, is now being applied to stars. It thus has been possible to locate the He II ionization feature at 55 $\pm$ 7% of the acoustic radius in the red Giant star HD 181907 [30] and at 809 $\pm$ 45s acoustic depth in the early F star HD 49933 [26]. The signature of the Base of the convective zone can also be addressed even if it is more delicate to asses [26].

Another insight into convection is brought by amplitudes and modes profile which are closely linked with convective velocities distribution and convection/oscillation interaction. First measurements with CoRoT on early F stars [1, 28] revealed amplitudes lower than expected by $\sim$25%, and modes lifetimes shorter (by a factor 2–3) than the present theoretical estimates (G. Houdek, private communication). Subsequent studies showed that this discrepancy could be explained partly, but partly only, by metallicity [37]. This investigation is being extended to hundreds of red giants observed with CoRoT and Kepler [38].They are found to show a similar trend for amplitudes, but very different for lifetimes [4].

Granulation is another manifestation of convection near the surface. Here again, first results on early F pulsators have revealed that a clear signature of granulation timescales and characteristic dimension could be obtained from the Fourier spectrum [28]. The comparison to 3D simulations revealed a strong disagreement, while the solar case was very satisfactorily reproduced [23]. This work is also being extended to red Giants observed during 13 months with Kepler [24].

Rotation: The light curve analysis of Main Sequence solar-like pulsators so far regularly provides an estimate of the average internal rotation. In many cases, it is also possible to derive an estimate of surface rotation from the spot rotational modulation. But only very recently, long runs provided by Kepler revealed individual frequency rotational splittings in red giant stars [6] with clear signature of non-rigid rotation. This exciting result should be followed rapidly by others.

Magnetic activity: Magnetic field is suspected to play a role in several aspects of stellar structure and evolution. New perspectives have been opened by the detection of the signature of an activity cycle in frequencies and linewidth variations for HD 49933 [18]. In addition, many light curves revealed the clear signature of spots, making spot modeling possible [22]. The level of precision in terms of reconstruction of the geometrical distribution of the spots is not as high as with Doppler imaging, but the large sample considered and the high sensitivity makes statistical studies promising.

Seismic indices: Beyond the unique opportunity to probe specific features in the interior of selected stars, seismology provides us with new observables to complement traditional ones for a large sample of stars. The interest of these new observables is being explored for Main Sequence and subgiant stars [13], but it is particularly well illustrated by the results obtained on red giants.

The very first long run of CoRoT revealed about 700 new pulsating red giants when less than ten were known before [14]. A couple of years after, thanks to CoRoT and Kepler, the red giants domain is one of the most advanced for seismology.

This is mainly due to the possibility to measure seismic indices associated with patterns characterizing the stellar envelope as in Main Sequence solar-type pulsators [7, 33], but also patterns associated with mixed-modes, which can be used to characterize the core [5, 32].

Several possible applications have been proposed already, as deriving seismic radius and mass estimates [20], or distinguishing between different evolutionary states (red clump vs red giant branch), while the stellar surface properties might be identical... [9]. These seismic indices have been used to compare an observed stellar population with a theoretical population synthesis and discuss the possible occurrence of recent stellar burst [29], but also to estimate mass loss rate at the tip of red giant branch [31].

5 Conclusions

Space photometry projects have kept their promises to collect light-curves of foreseen quality in terms of noise level, duration, duty cycle. To a great extent, the analysis of these time series is providing us with mode parameters determined with the anticipated level of precision. Finally, the first theoretical works conducted so far address most of the initial scientific objectives and they confirm that these data offer us a new grip on stellar physics and beyond on connected fields of research.

In addition, the analysis revealed that there is more than just oscillations to be found in photometry monitoring, with granulation, magnetic activity, binarity, outbursts,...

We have also shown that we can collect large amounts of light curves and analyse them to produce relevant seismic indices... This looks encouraging in the perspective of addressing larger samples and... why not... global sky surveys.

References

1. Appourchaux, T., Michel, E., Auvergne, M., et al.: A&A **488**, 705–714 (2008)
2. Auvergne, M., Bodin, P., Boisnard, L., et al.: A&A **506**, 411–424 (2009)
3. Baglin, A., Michel, E.: ArXiv e-prints (2011)

4. Baudin, F., Barban, C., Belkacem, K., et al.: A&A **529**, A84 (2011)
5. Beck, P.G., Bedding, T.R., Mosser, B., et al.: Science **332**, 205 (2011)
6. Beck, P.G., Montalban, J., Kallinger, T., et al.: Nat **481**, 55 (2012)
7. Bedding, T.R., Huber, D., Stello, D., et al.: ApJ **713**, L176–L181 (2010)
8. Bedding, T.R., Kjeldsen, H., Bouchy, F., et al.: A&A **432**, L43–L48 (2005)
9. Bedding, T.R., Mosser, B., Huber, D., et al.: Nature **471**, 608–611 (2011)
10. Benomar, O., Baudin, F., Campante, T.L., et al.: A&A **507**, L13–L16 (2009)
11. Benomar, O., Baudin, F., Marques, J.P., et al.: Astron. Nachr. **331**, 956 (2010)
12. Campante, T.L., Handberg, R., Mathur, S., et al.: A&A **534**, A6 (2011)
13. Chaplin, W.J., Kjeldsen, H., Christensen-Dalsgaard, J., et al.: Science **332**, 213 (2011)
14. De Ridder, J., Barban, C., Baudin, F., et al.: Nature **459**, 398–400 (2009)
15. Degroote, P., Aerts, C., Baglin, A., et al.: Nature **464**, 259–261 (2010)
16. Deheuvels, S., Bruntt, H., Michel, E., et al.: A&A **515**, A87 (2010)
17. Deheuvels, S., Michel, E.: A&A **535**, A91 (2011)
18. García, R.A., Mathur, S., Salabert, D., et al.: Science **329**, 1032 (2010)
19. Gruberbauer, M., Huber, D., Kuschnig, R., et al.: A&A **530**, A135 (2011)
20. Kallinger, T., Weiss, W.W., Barban, C., et al.: A&A **509**, A77 (2010)
21. Kjeldsen, H., Christensen-Dalsgaard, J., Handberg, R., et al.: Astron. Nachr. **331**, 966 (2010)
22. Lanza, A.F., Bonomo, A.S., Moutou, C., et al.: A&A **520**, A53 (2010)
23. Ludwig, H.G., Samadi, R., Steffen, M., et al.: A&A **506**, 167–173 (2009)
24. Mathur, S., Hekker, S., Trampedach, R., et al.: ApJ **741**, 119 (2011)
25. Matthews, J.M.: Comm. Asteroseismology **150**, 333 (2007)
26. Mazumdar, A., Michel, E.: ArXiv e-prints (2010)
27. Michel, E., Baglin, A., Auvergne, M., et al.: ESA Special Publication, vol. 1306, p. 39 (2006)
28. Michel, E., Baglin, A., Auvergne, M., et al.: Science **322**, 558 (2008)
29. Miglio, A., Montalbán, J., Baudin, F., et al.: A&A **503**, L21–L24 (2009)
30. Miglio, A., Montalbán, J., Carrier, F., et al.: A&A **520**, L6 (2010)
31. Miglio, A., Brogaard, K., Stello, D., et al.: MNRAS **419**, 2077 (2012)
32. Mosser, B., Barban, C., Montalbán, J., et al.: A&A **532**, A86 (2011)
33. Mosser, B., Belkacem, K., Goupil, M.J., et al.: A&A **525**, L9 (2011)
34. Poretti, E., Michel, E., Garrido, R., et al.: A&A **506**, 85–93 (2009)
35. Rowe, J.F., Matthews, J.M., Kuschnig, R., et al.: MmSAI **77**, 282 (2006)
36. Roxburgh, I.W.: ESA Special Publication, vol. 1306, p. 521 (2006)
37. Samadi, R., Ludwig, H.G., Belkacem, K., et al.: A&A **509**, A16 (2010)
38. Stello, D., Huber, D., Kallinger, T., et al.: ApJ **737**, L10 (2011)

Long-Baseline Interferometric Observations of Cepheids

P. Kervella, A. Mérand, and A. Gallenne

Abstract Although the period-luminosity relation was discovered 100 years ago, Cepheids are still surprising stars in many respects. The recent availability of optical/infrared long-baseline interferometers with hectometric baselines allows us to measure their changing angular diameters with sub-percent accuracy. Through a variant of the classical Baade-Wesselink method, interferometric measurements, complemented by spectroscopy, give us their distance and radius from the comparison of their angular and linear radius variation amplitudes. This elegant and accurate method, although quasi-geometrical, presents some particular difficulties that we briefly discuss. We also present the current status of our ongoing Cepheid observing program using the CHARA and VLTI interferometers.

1 The Interferometric Baade-Wesselink Method

Cepheids are certainly the most important class of variable stars in astronomy, thanks to the tight relation between their pulsation period and their intrinsic luminosity. Discovered more than 100 years ago [4, 5], they are still one of the key elements of the extragalactic distance scale, e.g. [10].

P. Kervella (✉)
LESIA, Observatoire de Paris, CNRS UMR 8109, UPMC, Université Paris Diderot, 5 place Jules Janssen, 92195 Meudon, France
e-mail: pierre.kervella@obspm.fr

A. Mérand
ESO, Alonso de Córdova 3107, Casilla 19001, Santiago 19, Chile
e-mail: amerand@eso.org

A. Gallenne
LESIA, Observatoire de Paris, France and ESO, Chile
e-mail: alexandre.gallenne@obspm.fr

Kervella, P. et al.: *Long-Baseline Interferometric Observations of Cepheids*.
Astrophys Space Sci Proc. **31**, 151–154 (2013)
DOI 10.1007/978-3-642-29630-7_29, © Springer-Verlag Berlin Heidelberg 2013

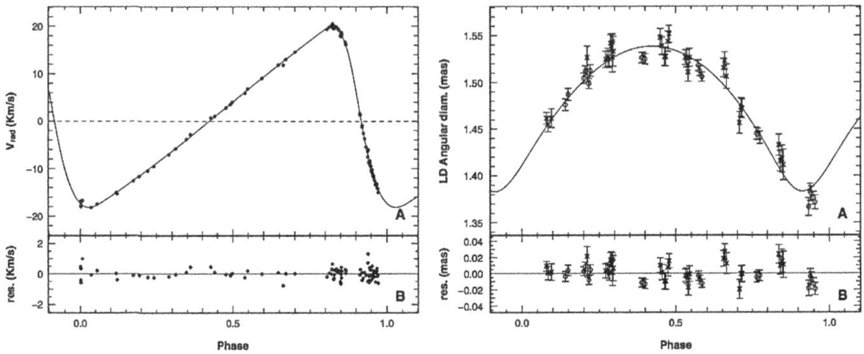

Fig. 1 Application of the interferometric Baade–Wesselink technique to the prototype Cepheid δ Cep (figure taken from [6]). *Left:* Radial velocity curve interpolated using periodic cubic splines. *Right:* Limb darkened disk angular diameter curve from interferometric measurements

The principle of the interferometric Baade-Wesselink (BW) method is to compare the variation in angular diameter of a pulsating star with the variation in linear radius derived from the integration of its radial velocity curve (Fig. 1). The equation that summarizes this technique is:

$$d = \frac{2\delta R(T)}{\delta\theta(T)} = \frac{-2\,kp\,\int_0^T v_{\rm rad}(t)\,dt}{\theta_{\rm UD}(T) - \theta_{\rm UD}(0)} \tag{1}$$

where d is the distance to the star, $\delta R(T)$ the variation in linear radius between epochs 0 and T, $v_{\rm rad}$ the radial velocity (as measured by spectroscopy, integrated over the apparent disk of the star), and $\theta_{\rm UD}$ the uniform disk equivalent angular diameter of the Cepheid. Two additional factors are also involved in this equation: the limb darkening (LD) coefficient $k = \theta_{\rm LD}/\theta_{\rm UD}$ and the projection factor p. The LD coefficient k is relatively easy to compute using classical stellar atmosphere models (e.g. ATLAS, PHOENIX,...), in particular at near-infrared wavelengths where the interferometric angular diameter measurements are usually obtained. The projection factor $p = v_{\rm puls}/v_{\rm rad}$ (p-factor) is more complex, as it relates the line-forming regions in the Cepheid atmosphere, the limb darkening at the wavelength of the spectroscopic data (usually in th visible), and the photosphere dynamics (its velocity is $v_{\rm puls}$). The p-factor is particularly important as a distance deduced from the application of any form (interferometric or classical) of the BW technique depends linearly on its value. For further details on the projection factor, the interested reader is refereed to the review by N. Nardetto in the present proceedings.

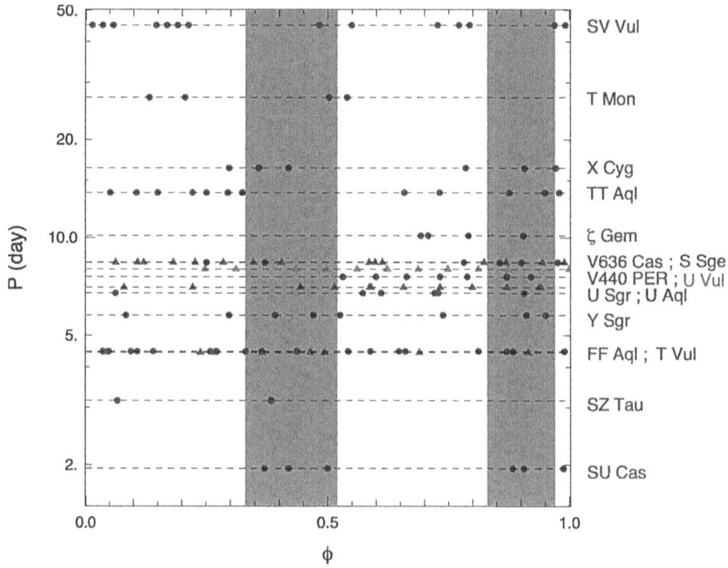

Fig. 2 Phase coverage of the sample of Cepheids observed with the CHARA array. The vertical axis gives the pulsation period of each Cepheid, in days

2 Status of Our Program

A few years ago, we embarked in a long-term program to monitor a sample of 20–30 nearby classical Cepheids using long-baseline interferometry at near-infrared wavelengths. Our final goal is the calibration of the Period-Luminosity (PL) relation, based on the application of the interferometric BW method, to obtain the zero point of the PL relation with an accuracy of the order of 0.01 mag. The results obtained on ℓ Car [1,2], δ Cep [6] and η Aql [8] show that the interferometric BW technique has the potential to provide Cepheid distances to a few percent accuracy. We therefore selected a sample of bright classical Cepheids for interferometric observations with the CHARA/FLUOR [12] and VLTI [13]. We obtained repeated angular diameter measurements covering the pulsation cycle of 24 stars in total, of which 22 are suitable for the application of the interferometric BW technique. Polaris (very low amplitude pulsations) and RS Pup (too small angular diameter for the VLTi baselines) are presently not accessible to measurements using this technique. Our phase coverage for the northern Cepheids is presented in Fig. 2, while the southern stars are listed in [1].

3 Conclusion

We are currently completing the interferometric observations of our reference sample of Galactic Cepheids. With 22 stars (with detectable angular diameter variations), completed by a few additional targets from the literature, this relatively large sample will soon provide us with an independent calibration of the zero point of the PL to a few percents. The application of the interferometric BW technique is complicated by the presence of circumstellar envelopes, e.g. [3,7], and uncertainties on the p-factor value [9, 11], but our observing procedure is now well established and will provide us with accurate and unbiased distances.

Acknowledgements This work received the support of PHASE, the high angular resolution partnership between ONERA, Observatoire de Paris, CNRS and University Denis Diderot Paris 7.

References

1. Kervella, P., Nardetto, N., Bersier, D., et al.: Cepheid distances from infrared long-baseline interferometry. I. VINCI/VLTI observations of seven Galactic Cepheid. A&A (2004) doi: 10.1051/0004-6361:20031743
2. Kervella, P., Fouqué, P., Storm, J., et al.: The angular size of the Cepheid ℓ Carinae: A comparison of the interferometric and surface brightness techniques. ApJ (2004) doi: 10.1086/383571
3. Kervella, P., Mérand, A., Perrin, G., Coudé du Foresto, V.: Extended envelopes around Galactic Cepheids. I. ℓ Carinae from near and mid-infrared interferometry with the VLTI. A&A (2006) doi: 10.1051/0004-6361:20053603
4. Leavitt, H.S.: 1777 variables in the magellanic clouds. Ann. Harvard College Observatory **60**, 87 (1908)
5. Leavitt, H.S., Pickering, E.C.: Periods of 25 variable stars in the small magellanic cloud. Harvard College Observatory Circular **173**, 1 (1912)
6. Mérand, A., Kervella, P., Coudé du Foresto, V., et al.: The projection factor of δ Cep. A&A (2005) doi: 10.1051/0004-6361:200500139
7. Mérand, A., Kervella, P., Coudé du Foresto, V., et al.: Extended envelopes around Galactic Cepheids. II. Polaris and δ Cephei from near-infrared interferometry with CHARA/FLUOR. A&A (2006) doi: 10.1051/0004-6361:20054466
8. Mérand, A., et al.: Interferometric observations of η Aql. A&A **532**, 50 (2012)
9. Nardetto, N., Fokin, A., Fouqué, P., et al.: The Baade-Wesselink p-factor applicable to LMC Cepheids. A&A (2011) doi: 10.1051/0004-6361/201117459
10. Riess, A.G., Macri, L., Casertano, S., et al.: A 3% Solution: Determination of the Hubble Constant with the Hubble Space Telescope and Wide Field Camera 3 (2011) doi: 10.1088/0004-637X/730/2/119
11. Storm, J., Gieren, W., Fouqué, P., et al.: Calibrating the Cepheid period-luminosity relation from the infrared surface brightness technique. I. The p-factor, the Milky Way relations, and a universal K-band relation. A&A (2011) doi: 10.1051/0004-6361/201117155
12. ten Brummelaar, T.A., McAlister, H.A., Ridgway, S.T., et al.: First results from the CHARA array. II. A description of the instrument. ApJ (2005) doi: 10.1086/430729
13. Wittkowski, M., Comerón, F., Glindemann, A., et al.: Observing with the ESO VLT interferometer. The Messenger (ISSN0722-6691), **119**, 14–17 (2005)

Programs and Perspectives of Visible Long Baseline Interferometry VEGA/CHARA

D. Mourard, N. Nardetto, R. Ligi, and K. Perraut

Abstract VEGA/CHARA is a visible spectro-interferometer installed on the CHARA Array at Mount Wilson Observatory. Combining high spectral resolution (6,000 or 30,000) and high angular resolution (0.3 mas), VEGA/CHARA opens a wide class of astrophysical topics in the stellar physics domain. Circumstellar environments and fundamental parameters with a high precision could be studied. We will present a review of recent results and discuss the programs currently engaged in the field of pulsating stars and more generally for the fundamental stellar parameters. Details could be found at http://www-n.oca.eu/vega/en/publications/index.htm.

1 Why Measuring Angular Diameters?

Optical long baseline interferometry is now a mature observing technique and large worldwide facilities like VLTI [1] and CHARA [2] offer unique high angular resolution capabilities at wavelengths between 0.5 and 10 μm.

High angular resolution is mainly used for circumstellar studies as well as surface imaging of stars. Astrometric performance allows also studying multiple systems or binaries like star-planet couples. With the very long baselines (more than 300 m) of the CHARA Array and the short wavelength accessible to the VEGA instrument [3, 4], an angular resolution as small as 0.3 millisecond of arc is now accessible. It allows measuring angular diameters over a very wide range of

D. Mourard (✉) · N. Nardetto · R. Ligi
Laboratoire Fizeau, UNS/OCA/CNRS UMR6525, BP 4229, 06304 Nice Cedex, France
e-mail: denis.mourard@oca.eu

K. Perraut
IPAG, UJF/CNRS, 414, rue de la Piscine, Domaine Universitaire,
38400 Saint-Martin d'Hères, France

Mourard, D. et al.: *Programs and Perspectives of Visible Long Baseline Interferometry VEGA/CHARA*. Astrophys Space Sci Proc. **31**, 155–158 (2013)
DOI 10.1007/978-3-642-29630-7_30,

stellar classes and gives access to stellar fundamental parameters complementary of other observing techniques. When measuring angular diameters, one can use simple stellar laws to calibrate for example surface brightness relations that are important for distance determination or deduce constraints on stellar atmosphere models, like effective temperature for example. With the addition of a known and accurate parallax, it is possible to determine a linear radius that will bring strong constraints on stellar evolution models (mass, age, ...) or on models developed for asteroseismology frequencies understanding. A review of the interest of measuring angular diameter in complement to asteroseismology studies could be found in [5] or [6].

2 What Is VEGA/CHARA?

VEGA/CHARA has been described in details in two recent papers [3, 4]. The VEGA spectrograph is designed to sample the visible band from 0.45 to 0.85 μm with spectral resolution of 1,700, 6,000, and up to 30,000. It is equipped with two photon counting detectors that simultaneously study two different spectral bands at high and medium spectral resolutions. The main characteristics are summarized in Table 1. The simultaneous operation of the two detectors is only possible in high and medium spectral resolution. Observing in the blue requires good seeing conditions but increases by 30% the limit of spatial resolution of the instrument with respect to its operation around 700 nm.

We indicate in Table 2 the typical limiting magnitudes as well as the best performance obtained in the different observing modes.

The most recent results related to this conference are the ones recently published by [7] or [8]. We observed γ Equ with VEGA and derived both the uniform-disk angular diameter and the limb-darkened diameter from the calibrated squared visibility. We then determined the luminosity and the effective temperature of the star from the whole energy flux distribution, the parallax, and the angular diameter. Thanks to the unique capabilities of VEGA, we managed to constrain the angular diameter of a star as small as 0.564 mas with an accuracy of about 3% and to derive its fundamental parameters. In particular the new values of the radius and effective temperature should bring further constraints on the asteroseismic modeling of the star. In fact such Ap stars are known to present spotty surfaces, which induces bias in the determination of effective temperature by photometric methods. Deriving direct angular diameters for these stars is a strong mean to put into test stellar interior models. Regarding HD49933 (one of the primary target of the CoRoT mission), we use the direct interferometric determination of the angular diameter and advanced three-dimensional (3D) modeling for reducing the global stellar parameter space compatible with seismic data. We obtain a limb-darkened angular diameter of ϕ_{LD} = 0.445±0.012 mas. Combined with the Hipparcos parallax, we obtain a radius of R = 1.42 ± 0.04 R_{sun}.

Table 1 Spectral resolution (R) and bandwidth ($\Delta\lambda$) of the VEGA spectrograph, as well as the spectral separation between the two detectors

Grating	R	$\Delta\lambda$ (Blue)	$\Delta\lambda$ (Red)	$\lambda_R - \lambda_B$
R1: 1,800 gr/mm	30,000	5 nm	8 nm	25 nm
R2: 300 gr/mm	5,000	30 nm	45 nm	170 nm
R3: 100 gr/mm	1,700	100 nm	150 nm	Not possible

Table 2 Estimation of typical limiting magnitude as a function of the different spectral resolution modes. These values are presented for the median value of r0 at Mount Wilson i.e. 8 cm. We also indicate the best performances assuming an r0 of 15 cm.

Resolution	R	Typical Lim. Magnitude	Best perf.
Low	1,700	6.8	7.5
Medium	6,000	6.5	7.5
High	30,000	4.2	5.5

3 Science Perspectives of Very Small Stellar Angular Diameter Measurements

Thanks to *direct* angular diameter measurements by optical interferometry combined with the bolometric flux, it is possible to derive the fundamental stellar parameters, such as the effective temperature, the luminosity, and the stellar radius. With VEGA, being able to measure very low visibilities will allow us to determine the angular diameters of stars of sub-mas sizes to an accuracy better than 2%. In order to determine limb darkening laws, we limit ourselves to stars presenting an angular diameter smaller than 3.0 mas in order to be able to cover both the central lobe and the side lobes of the visibility function. Moreover, using triplets of the longest baselines in the visible is a powerful means of determining accurate limb darkening laws throughout the HR diagram. Limb darkening is a function of the wavelength, and in particular limb darkening in a given spectral line depends upon the physical location in the atmosphere of the line-forming region of the element. Thus, an accurate measurement of the limb-darkening within spectral lines will allow us to study the temperature and density structure of the atmosphere, hence the strong interest in using multi-telescope combiners operating with dispersed fringes.

For VEGA, by adopting the same approach used for searching to calibrators and considering the standard VEGA squared visibility measurement accuracy of 2% and the CHARA longest baseline (330 m), we have determined the minimum measurable diameter for a given accuracy. This corresponds to 0.16 mas (resp. 0.12, 0.075) for a 1% (resp. 2%, 5%) accuracy in the angular diameter determination. We have then determined that such limb-darkening studies will be possible for stars spread over nearly the whole Hertzsprung-Russell diagram (excluding compact

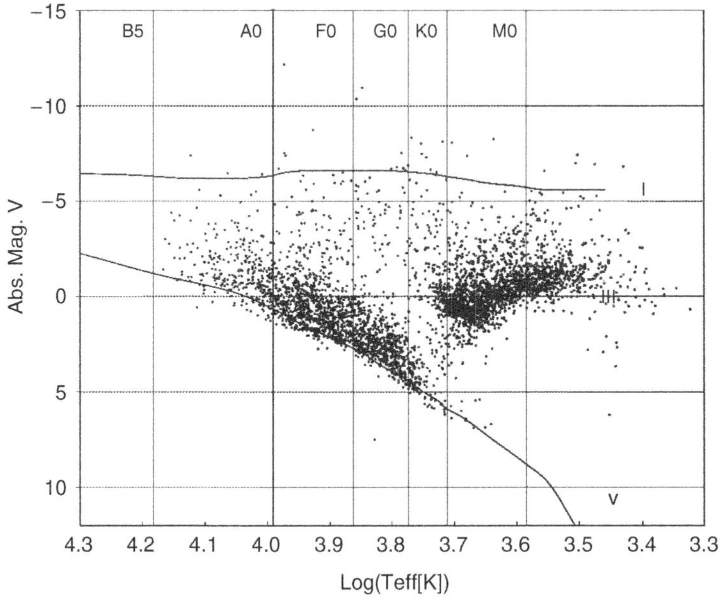

Fig. 1 Sample of stars brighter than $m_V = 6.5$ and for which VEGA can determine the angular diameter with a precision better than 2%

objects) as illustrated in Fig. 1, and will result in a very significant improvement in the stellar atmosphere models (1D and 3D) of all stars.

References

1. Glindemann, A., Albertsen, M., Andolfato, L., et al.: SPIE **5491**, 447 (2004)
2. ten Brummelaar, T.A., McAlister, H.A., Ridgway, S.T., et al.: ApJ **628**, 453–465 (2005)
3. Mourard, D., Clausse, J.M., Marcotto, A., et al.: A&A **508**, 1073 (2009)
4. Mourard, D., Bério, Ph., Perraut, K., et al.: A&A **531**, 110 (2011)
5. Cunha, M.S., Aerts, C., Christensen-Dalsgaard, J., et al.: A&A Rev. **14**, 217 (2007)
6. Creevey, O.L., Monteiro, M.J.P.F.G., Metcalfe, T.S., et al.: ApJ **659**, 616 (2007)
7. Perraut, K., Brandão, I., Mourard, D., et al.: A&A **526**, 89 (2011)
8. Bigot, L., Mourard, D., Berio, P.: A&A **534**, 3 (2011)

Spectro-Interferometry Studies of Velocity-Related Phenomena at the Surface of Stars: Pulsation and Rotation

Antoine Mérand, Fabien Patru and Jason Aufdenberg

Abstract We illustrate here two applications of spectro-interferometry to the study of velocity fields at the surface of stars: pulsation and rotation. Stellar pulsation has been resolved spectroscopically for a long time, and interferometry has resolved stellar diameters variations due to pulsation. Combining the two provides unique insights to the study of Cepheids, in particular regarding the structure of the photosphere or investigating the infamous projection factor which biases distances measured by the Baade–Wesselink method. On the other hand, resolving the surface velocity field of rotating stars offers a unique opportunity to potentially study differential rotation in other cases than for the Sun.

We also present the model we have implemented recently, as well as two applications to VLTI/AMBER Data: the pulsation of Cepheids and the rotation of intermediate mass main sequence stars.

1 Rotation of Main Sequence Stars

Rotation is an important factor that deeply influences the internal structure of a star. Surface rotation can be studied using spectroscopy (Doppler broadening of absorption and/or emission lines), the flux modulation due to dark spots (Doppler imaging) or the line profile variation due to an occultant (a.k.a Rossiter-McLaughlin effect). Stellar interferometry has already provided reconstructed center-to-limb profiles [1] and images of the surface of rotating stars [2, 7, 8, 12]. The results of

A. Mérand (✉) · F. Patru
European Southern Observatory (ESO), Alonso de Cordova 3107, Vitacura Santiago, Chile
e-mail: amerand@eso.org

J. Aufdenberg
Embry-Riddle Aeronautical University 600 S. Clyde Morris Blvd. Daytona Beach
FL 32114, USA

Mérand, A. et al.: *Spectro-Interferometry Studies of Velocity-Related Phenomena at the Surface of Stars: Pulsation and Rotation*. Astrophys Space Sci Proc. **31**, 159–163 (2013)
DOI 10.1007/978-3-642-29630-7_31,

these studies derived from our simple model using a Roche model for the shape of the uniformly (latitudinally) rotating star, Von-Zeipel gravity darkening (locally, $T_{\rm eff} \propto g^{\beta}$, with classically $\beta = 0.25$) and the mapping of atmospheric models computed for non-rotating stars, for each element at the surface of the star (see [1] for a more detailed description).

In these studies, the model has been used to derive the physical parameters of the star using broadband interferometric observations: the only parameters of the model are M, R, L, $\omega/\omega_{\rm crit.}$, i, PA and β (Mass, polar radius, luminosity, fractional rotational rate, orientation angles and gravity darkening). More elaborated models are being developed, e.g. [3] but they do not yet output interferometric observables so their comparison to interferometric data is not straightforward. Ultimately, the goals of this study are: (1) demonstrate the feasibility of deriving physical parameters from spectro-interferometric data inside an absorption line, (2) verify the agreement with broad-band interferometric data, and (3) study the possibility to derive additional information, such as differential rotation.

Altair is an A7V rotating star which has been extensively studied, including by stellar interferometry. We present here archival data obtained with VLTI/AMBER,[1] at a spectral resolution of $R \sim 15,000$ around the hydrogen Brackett γ line ($\lambda \approx 2.166\,\mu$m). We model the interferometric observables using the same approach as described before, updating the numerical code to map spherical PHOENIX [4] models, Doppler shifted according to the projected rotational velocity of each surface element. Since AMBER resolves the line spectroscopically *and* spatially, the signal depends on the orientation and the length of the baseline. Even without a fit, the agreement of our data with others [7] is very good within the framework of this modeling, showing encouraging results (see Fig. 1).

2 Cepheids' Pulsations

Cepheids' pulsations have been used for a very long time in order to derive distances. It is of prime importance, for example, to derive observationally the Hubble constant H_0 [11]. Our understanding of the physical processes underlying the pulsation process is important for two reasons: (1) we need to make sure we can justify the pragmatic approach of using Cepheids' Period–Luminosity (P–L) relation in any galaxy, for any period range etc. without introducing any bias and (2) we need to understand the projection factor [6, 10] in order to calibrate the parallax of pulsation method (a.k.a Baade–Wesselink method).

The preliminary analysis of our VLTI/AMBER data of the Cepheid ℓ Car show that the spectrum is well reproduced by a PHOENIX model which has been properly modified to take into account the center-to-limb Doppler shift from the pulsation (Fig. 2). Moreover, for the largest baseline of this particular data set ($B \approx 126$ m),

[1] http://www.eso.org/sci/facilities/paranal/instruments/amber/

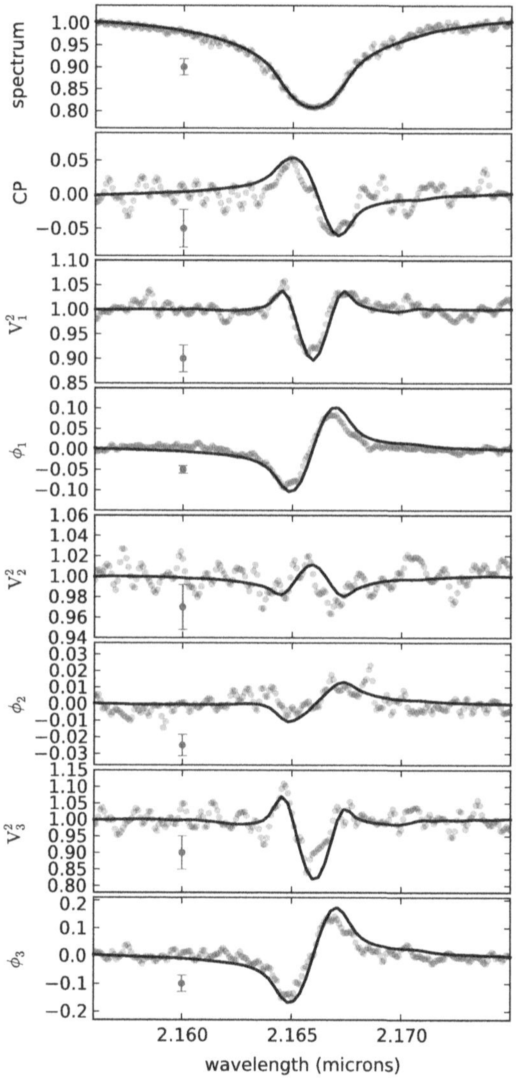

Fig. 1 AMBER Observations of the fast rotating star Altair. From top to bottom, as a function of wavelength, normalized spectrum, differential phase closure (CP, in rad) then for each baseline ($i = 1...3$), differential squared visibilities (V_i^2) and differential phases (ϕ_i, in radian). The *scattered dots* are the data and the *fine line* corresponds to the model using parameters from Monnier et al. [7], based on their fit of broad band data using the CHARA/MIRC interferometer: $M = 1.791\, M_{\odot}$, apparent $R_{\rm pol} = 1.479$ mas, $T_{\rm pol} = 8,450$ K, apparent $R_{\rm eq} = 1.835$ mas, $T_{\rm eq} = 6,860$ K, $\omega/\omega_{\rm crit} = 0.923$, $\beta = 0.190$, $i = 57.2^{\rm o}$, $PA = -61.8^{\rm o}$. The agreement is very good, even thought this is not a fit. This confirms the consistency of the VLTI/AMBER and CHARA/MIRC data within the models we use. Note that these models, though independent implementations, are very similar in principle and have been cross validated

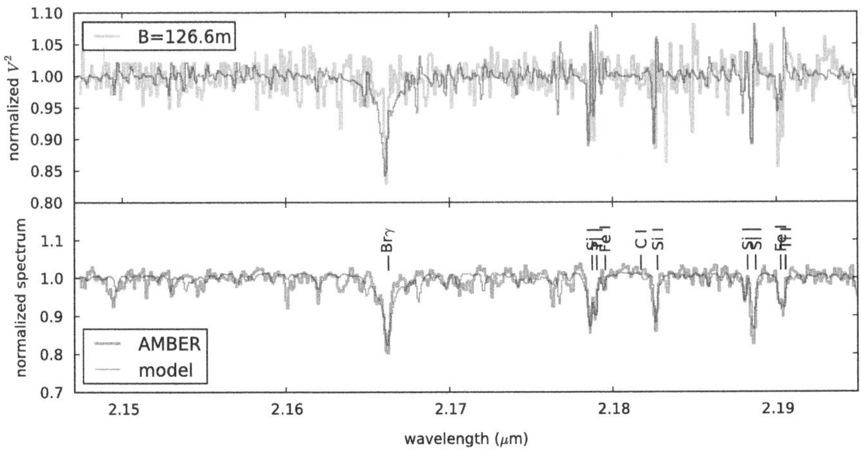

Fig. 2 AMBER Observations of ℓ Car at pulsation phase $\phi \approx 0.64$. The *upper plot* is the differential squared visibility (normalized to the continuum), the *lower panel* is the normalized spectrum as a function of wavelength. The *thick colored lines* are the observed data, the *fine line* is the current prediction of our model using the expected values for $T_{\rm eff}$, $logg$, angular diameter θ and pulsation velocity (based on [5]). The *metallic lines* have the expected [9] "S" shape in the visibility

we detect two kinds of signals: a differential visibility deficit for the Brackett γ line and the expected "S" like features for the metallic lines [9]. The data, though noisy, are promising and require further analysis.

3 Perspectives

We have shown two examples of ongoing work to resolve velocity fields at the surface of stars using spectro-interferometry with VLTI/AMBER: rotation and pulsation. In both cases, we resolve the stellar surface in absorption lines and a signal is detected in differential phase and/or visibilities. This opens new perspectives in these two fields, for example to the study of latitudinal differential rotation for the star Altair and the direct derivation of the projection factor for the Cepheid ℓ Car.

References

1. Aufdenberg, J., Mérand, A., Coudé du Foresto, V., et al.: ApJ **645**, 664 (2006)
2. Che, X., Monnier, J.D., Zhao, M., et al.: ApJ **732**, 68 (2011)
3. Espinosa Lara, F., Rieutord, M.: A&A **533**, 43 (2011)
4. Hauschildt, P.H., Allard, F., Ferguson, J., Baron, E., Alexander, D.R.: ApJ **525**, 871 (1999)

5. Kervella, P., Nardetto, N., Bersier, D., Mourard, D., Coudé du Foresto, V.: A&A **416**, 941 (2004)
6. Mérand, A., Kervella, P., Coudé du Foresto, V., et al.: A&A **438**, 9 (2005)
7. Monnier, J.D., Zhao, M., Pedretti, E., et al.: Science **317**, 342 (2007)
8. Monnier, J.D., Townsend, R.H.D., Che, X.: ApJ **725**, 1192 (2010)
9. Nardetto, N., Fokin, A., Mourard, D., Mathias, Ph.: A&A **454**, 327 (2006)
10. Nardetto, N., Mourard, D., Mathias, Ph., et al.: A&A **471**, 661 (2007)
11. Riess, A.G., Macri, L., Castertano, S.: ApJ **730**, 119 (2011)
12. Zhao, M., Monnier, J.D., Pedretti, E.: ApJ **701**, 209 (2009)

The VISTA Variables in the Vía Láctea Survey: A First Glance on Stellar Variability

I. Dékány, M. Catelan, D. Minniti, and the VVV Collaboration

Abstract VISTA Variables in the Vía Láctea (VVV) is an ESO public near-infrared variability survey of the Galactic bulge and an adjacent area of the southern mid-plane. It will produce a deep atlas in the $ZYJHK$s filters, and a K_s-band time-series database of $\sim 10^9$ point sources, among which $\gtrsim 10^6$ are expected to be variable. One of VVV's immediate scientific goals is to provide accurate light curves of primary distance indicators, such as RR Lyrae stars, and utilize these data to produce a 3-D map of the surveyed area and, ultimately, trace the structure of the inner Galaxy. We give, based on the first $\sim$ 1.5 years of the 5-year-long survey, an early assessment on the basic properties and the overall quality of the VVV photometric time-series, and use these data to put an estimate of the fraction of variable stellar sources.

1 Introduction

VISTA Variables in the Vía Láctea (VVV) is an ESO Public Survey[1] of the Milky Way bulge and a significant area of the southern mid-plane [7]. It is carried out with VIRCAM [4] on the 4-m VISTA telescope in the $ZYJHK_s$ near-IR filter system, enabling us to see through low-latitude Galactic regions highly obscured by interstellar dust, and to probe deeper than ever into our galaxy, and beyond. The deep color map provided by VVV, covering 520 deg^2, and including 39 known globular and >400 open clusters, is already a treasure trove for various fields of astronomy,

[1]http://vvvsurvey.org/

I. Dékány (✉) · M. Catelan · D. Minniti
Pontificia Universidad Católica de Chile, Facultad de Física
Departamento de Astronomía y Astrofísica, Av. Vicuña Mackenna 4860, 782-0436 Macul, Santiago, CHILE

The VVV Collaboration
http://mwm.astro.puc.cl/mw/index.php/Main_Page

Dékány, I. et al.: *The VISTA Variables in the Vía Láctea Survey: A First Glance on Stellar Variability*. Astrophys Space Sci Proc. **31**, 165–169 (2013)
DOI 10.1007/978-3-642-29630-7_32,

and has already led to a series of important discoveries [1, 6, 8]. In addition, VVV is also exploring the time-domain by monitoring the complete surveyed area in the K_s band, enabling us not only to study time-varying phenomena, but also to produce a 3-D map of our galaxy using well-understood primary distance indicators, such as "classical" radial-mode pulsating stars. In the following, we give a short assessment on the expected properties of the time-series data being collected by our survey.

2 VVV Time-Series Photometry

The VVV Survey is performing repeated observations of its entire area in the K_s band. The 1st year was mainly devoted to the multi-color mapping of the bulge and the mid-plane. In the following years, the focus is put on the extension of the time series. At the time of writing, VVV is finishing its 2nd season, with up to 13 epochs per source already observed. The main variability campaign of the bulge and the disk will be performed in the course of the 3rd and 4th years, respectively, and a total of $\sim$ 100 epochs are scheduled to be obtained for $\sim 10^9$ point sources until the end of the 5th year.

Raw image data are transferred into the VISTA Data Flow System (VDFS) [5] for pipeline processing, consisting of image reduction, source extraction, aperture photometry, photometric calibration using 2MASS secondary standard stars, and digital curation. Final survey products are stored in a synoptic relational database [3], and can be accessed through the ESO Archives. Efforts are being made to incorporate additional procedures in the VDFS, such as PSF and image subtraction photometry, and automatic light curve classification[2] [2].

The limiting magnitude of the VVV survey is up to 4^m fainter than 2MASS, reaching $K_s = 18^m$ in some areas, although source confusion and the heavy interstellar extinction become severe limiting factors towards the Galactic center [10]. In general, VVV data will enable us to detect RR Lyrae stars out to the far edge of the Milky Way, whereas our horizon for G dwarfs will expand to as far as $\sim$ 8 kpc.

The sampling of VVV light curves should be nearly random, providing a homogenous phase coverage for variable stars such as pulsators in the classical instability strip and eclipsing binaries. Observations with high cadence (i.e., 10–40 times per night) are also planned to be carried out in some selected areas during the 5th year, in order to search for short-period variables and transient events. At each epoch, six consecutive offset exposures ("pawprints") are taken, in order to fill the gaps between the 16 chips of VIRCAM, and provide a mosaic image with contiguous sky coverage (a "tile"). Both tiles and pawprints are fully processed by the VDFS. An example VVV light curve taken from the VISTA Science Archives is shown in Fig. 1. The rms scatter of $\sim 2 \times 10^5$ stellar point sources in one of the most crowded areas of the bulge close to Baade's window is shown in Fig. 2a.

[2]http://www.vvvtemplates.org/

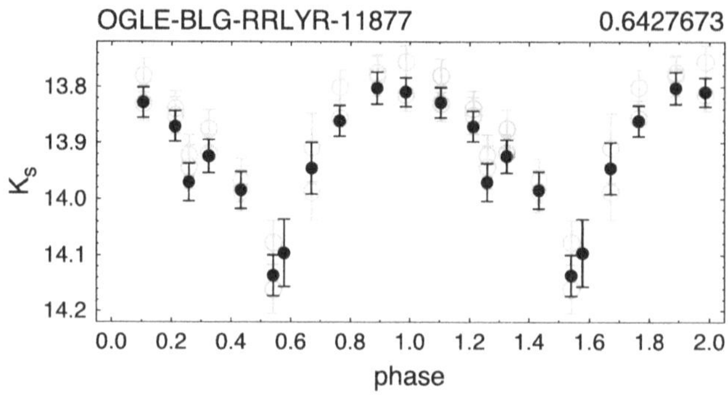

Fig. 1 Phase-folded K_s light curve of a known RRab Lyrae star from the OGLE3 survey [11]. *Gray and black symbols* denote pawprint and tile photometry, respectively. The identifier and period of the star (in days) are shown in the figure's header

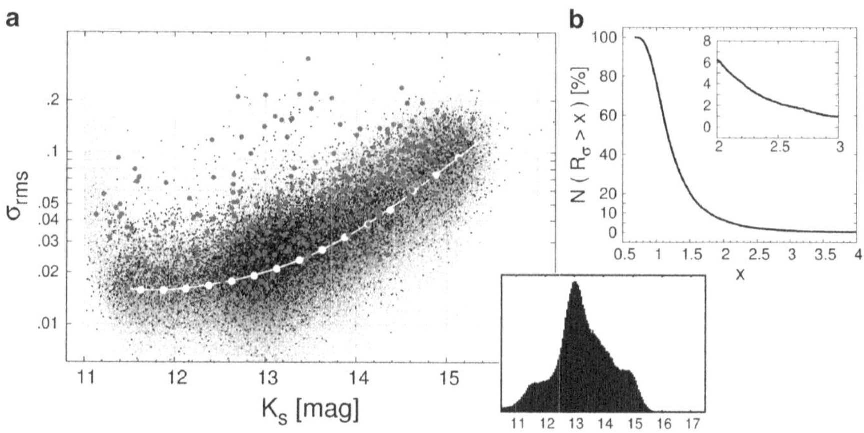

Fig. 2 (**a**): Scatter plot of stellar point sources in the bulge *(gray dots)*, with the inset showing the histogram of their K_s magnitudes. The *white curve* is a polynomial fitted to the binned values *(white points)*. Stars included in the variability search are shown by *black dots*, and those having $R_\sigma \geq 2.5$ are marked by *red points*. (**b**): Cumulative distribution of R_σ values of the objects shown in black in Fig. 2a. The inset shows a blowup of the figure around $x = 2.5$

The increased source density around $13^{\rm m}$ is due to red clump giants, while most of the stars below $15^{\rm m}.5$ do not yet have enough good detections to be shown. At any given magnitude, σ shows a large scatter, due to different levels of crowding. For comparison, Fig. 3 shows a similar figure for a much less crowded area in the disk, also less affected by extinction, resulting in a significantly higher photometric precision and limiting magnitude. (To some extent, the difference between these areas may also be partly due to seeing variations, since the onset of saturation is also different between them.)

In order to put an estimate on the fraction of likely variable stellar sources, we performed the following analysis. As a result of the acquisition method of VVV

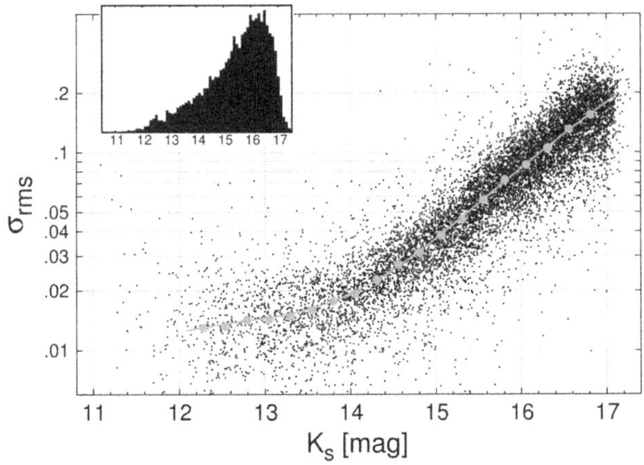

Fig. 3 Scatter plot and histogram *(inset)* of $\sim 10^4$ stellar point sources in two areas of the Galactic disk close to $l \simeq 310$, $b \simeq -2$ where VVV tiles overlap. The notation is similar to that of Fig. 2

images described above, there are up to 6 pawprint detections of a source within a few minutes, carrying information about the σ_{i} internal scatter of the time-series. We estimate this for each source having N pawprint detections by the quantity $\sigma_{\mathrm{i}} = \sum_{j=1}^{\mathrm{N}-1}(m_j - m_{j+1})^2/2\mathrm{M}$, where m_j is the magnitude at the j-th detection. The sum is evaluated for M pairs of detections, each made within a certain tile acquisition sequence. Various levels of σ-clipping were also performed in order to prevent the results from being dominated by outliers. The cumulative distribution of the R_σ ratio between the global rms scatter and σ_{i} is shown in Fig. 2b for $\sim 2\times10^4$ uncrowded bulge point sources with $\mathrm{M} \geq 8$ and low photometric error bits. Candidate variable stars with $R_\sigma \geq 2.5$ are shown with red points in Fig. 2a. Many of these clearly cannot be identified by simply estimating σ_{i} by a global fit to the σ_{rms}—magnitude distribution. The estimated fraction of ~1–2% variable sources is consistent with the results of [9], based on a variability search in the VVV area with VLT/VIMOS.

3 Summary

The ultimate goal of the VVV ESO Public Survey is to improve our global understanding of our galaxy's origin, evolution, and present structure. It will provide a deep near-IR color atlas and accurate K_s light curves for about a million variable stars, greatly expanding our present horizon and enabling us to produce a precise 3-D map of the Milky Way. VVV data combined with other surveys covering the whole electromagnetic spectrum will help us understand the physics of variable sources, and several important implications will also be drawn for stellar evolution, pulsation theory, and galaxy dynamics.

Acknowledgements The authors acknowledge the support of Programa Iniciativa Científica Milenio through grant P07-021-F, awarded to The Milky Way Millennium Nucleus; Proyecto Fondecyt #1110326; FONDAP Centro de Astrofísica 15010003; and Proyecto Basal PFB-06/2007 (CATA).

References

1. Borissova, J., Bonatto, C., Kurtev, R., et al.: A&A **532**, 131 (2011)
2. Catelan, M., Minniti, D., Lucas, P.W., et al.: Carnegie Obs. Astr. Ser. **5**, 145 (2011)
3. Cross, N.J.G., Collins, R.S., Hambly, N.C., et al.: MNRAS **399**, 1730 (2009)
4. Dalton, G.B., Caldwell, M., Ward, A.K., et al.: SPIE **6269**, 30 (2006)
5. Irwin, M.J., Lewis, J., Hodgkin, S., et al.: SPIE **5493**, 411 (2004)
6. Minniti, D., Hempel, M., Toledo, I., et al.: A&A **527**, 81 (2011)
7. Minniti, D., Lucas, P.W., Emerson, J.P., et al.: New Astron. **15**, 433 (2010)
8. Minniti, D., Saito, R.K., Alonso-García, J., et. al.: ApJL **733**, L43 (2011)
9. Pietrukowicz, P., Minniti, D., Alonso-García, J., Hempel, M. (2011) arXiv:1110.3453
10. Saito, R., Hempel, M., Alonso-García, J., et al.: The Messenger, 141, 24 (2010)
11. Soszyński, I., Dziembowski, W.A., Udalski, A., et al.: **AcA**, 61, 1 (2011)

Solar-Like Stars Observed by Kepler: An Incredible Adventure

R.A. García

Abstract The NASA *Kepler mission*—in flight since March 2009—is producing an enormous number of high-quality continuous light curves. Now, and for the first time ever, we are able to do ensemble asteroseismology, i.e., to do an asteroseismic analysis with a statistically significant sub-sample of solar-like stars covering a wide range of stellar characteristics. In the present work, I highlight some of the most recent studies carried out using these data.

1 Introduction

Kepler [7] is a planet-hunter mission to look for Earth-size planets around solar-like stars in the habitable zone [2]. The photometric stability and the quiet environment provided by the Earth trailing heliocentric orbit offers a great opportunity to perform asteroseismic observations of thousands of stars, after some specific processing of the light curves [18], among the 150,000 observed by *Kepler* at a single star field in the Cygnus-Lyra region of the galaxy. *Kepler* is planned to work for 3.5 years with an extension to 6 years. With its telescope of 0.95-m aperture, *Kepler* monitored 100,000 stars at any time with a cadence of 29.42 min (long-cadence measurements). A subsample of 512 stars can be observed at a much faster cadence of 58.85s (short-cadence measurements; Nyquist frequency of ~8.5 mHz) allowing for more precise transit timings [20]. This running mode is of particular interest for asteroseismology because it allows us to study stochastically-excited oscillations in main-sequence solar-like stars and subgiants [5, 10]. During the first year of operations, the working group 1 of the Kepler Asteroseismic Science

R.A. García on behalf of the WG#1 Team (✉)
Laboratoire AIM, CEA/DSM-CNRS-Université Paris Diderot; IRFU/SAp, Centre de Saclay, 91191, Gif-sur-Yvette, France
e-mail: rgarcia@cea.fr

García, R.A.: *Solar-Like Stars Observed by Kepler: An Incredible Adventure.*
Astrophys Space Sci Proc. **31**, 171–175 (2013)
DOI 10.1007/978-3-642-29630-7_33,

Consortium (KASC http://astro.phys.au.dk/KASC/) observed more than 2000 solar-like stars looking for stellar pulsations with 1-month long time series reporting their existence in more than 550 stars [13]. From the second year on, around 110 stars have been continuously observed for a year and additional ~190 stars have been observed for at least 3 continuous months. All this observational set constitutes an ensemble in which we can perform very accurate asteroseismic analyses. Moreover, stellar evolution studies of solar-like stars are completed by the analysis of several thousands of red giants measured in the *Kepler* long-cadence (e.g. [5,24]).

2 Some initial results

Asteroseismology is a powerful tool to look inside the stars and put strong constraints in structure and evolution models (e.g. [25,30,31]). In pulsating stars that host planets, asteroseismology allows us to also place very tight constraints on the exoplanetary systems [1,14].

Depending on the signal-to-noise ratio of the modes—which also seems to depend on the magnetic activity of the star [11]—on one hand we can have access to only some global seismic parameters (e.g. $\Delta\nu$, $\nu_{\rm max}$, $A_{\rm max}$ [36]) of the stars using a new generation of automatic procedures that have been set up for solar-like pulsating stars (e.g. [21, 23, 27, 32]) including red giants [22]. From these measurements we were able, for example, to verify [35] the $\log g$ and the radius of the *Kepler* Input Catalog [8] inferred from the scaling relations based on solar values ([12] and [26]). On the other hand—when the signal-to-noise ratio is high enough—we can characterize very precisely the individual p-mode parameters (e.g. [9,29]). In Fig. 1 the échelle diagram of KIC 11395018—observed for more than 8 continuous months—is shown. We clearly identify the ridges of the modes l=2, 0, and 1 from left to right respectively. A nice feature of the power spectrum is the presence of a mixed l=1 mode at 740.29 μHz that bumps the modes in the surrounding orders. Indeed mixed modes are very interesting. They behave like acoustic modes near the surface of the stars and as g modes in the interior. They have a very high sensitivity to the core of the stars. In the case of the red giants, many mixed modes can be measured (see e.g. [3]). They can be used to determine the status of the core (if they are already burning He or not [6]) and even to track the internal differential rotation ([4], [16]). In the case of the subgiants, where only a few mixed modes are available, their position in frequency depends very much on the age of the star and can be used to further constrain the models [15].

The high number of stars available to the asteroseismic research allows us to build evolutionary sequences of stars—of similar masses and compositions—and study the evolution of the physical processes governing their interiors as a function of time, like a collection of snapshots of the life of a star (see [34] for more details).

With the extension of the *Kepler mission* to 6 years, the future of the solar-like studies with *Kepler* is amazing. In particular, having several years of continuous observations of a several dozens of stars will allow us to study their dynamics:

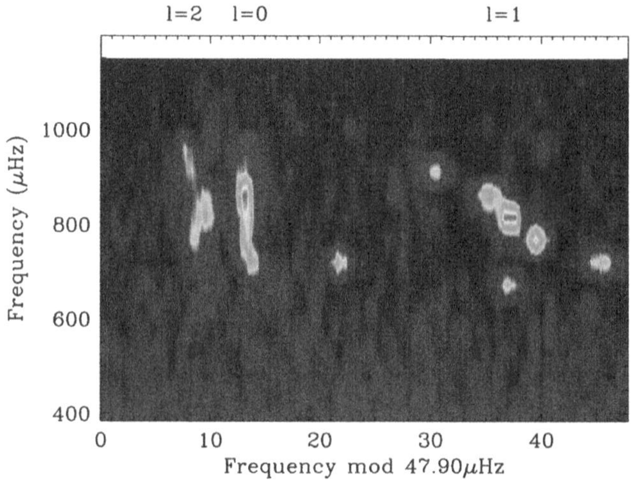

Fig. 1 Échelle diagram of the post main-sequence star KIC 11395018 showing a mixed mode between the ridges of the l=0 and 1 modes and a few bumped l=1 modes. Adapted from [29]

rotation [19] and magnetic fields. Indeed we will be able to study the surface magnetic activity [28] at the same time as the internal one [17], and the internal structure including the characteristics of the convective zones. The physical processes governing the dynamos will be better constrained and we will improve our knowledge of the solar dynamo, which still hides its secrets (e.g. [33]).

Acknowledgements RAG wishes to thank the *Kepler* team. Funding for this Discovery mission is provided by NASAs Science Mission Directorate. This work has received funding from the FP7/2007-2013 program No. 269194 and the NSF Grant No. NSF PHY05-51164.

References

1. Ballot, J., Gizon, L., Samadi, R., et al.: Accurate p-mode measurements of the G0V metal-rich CoRoT target HD 52265. A&A **530**, 97–108 (2011)
2. Batalha, N.M., Borucki, W.J., Bryson, S.T., et al.: *Kepler's* first rocky planet: Kepler-10b. ApJ **729**, 27–48 (2011)
3. Beck, P.G., Bedding, T.R., Mosser, B., et al.: *Kepler* detected gravity-mode period spacings in a red giant star. Science **332**, 205 (2011)
4. Beck, P.G., Montalban, J., Kallinger, T., et al.: Fast core rotation in red giant stars as revealed by gravity-dominated mixed modes. Nature **481**, 55–57 (2012)
5. Bedding, T.R., Huber, D., Stello, D., et al.: Solar-like oscillations in low-luminosity red giants: first results from *Kepler*. ApJ **713**, L176–L181 (2010)
6. Bedding, T.R., Mosser, B., Huber, D., et al.: Gravity modes as a way to distinguish between hydrogen- and helium-burning red giant stars. Nature **471**, 608–611 (2011)
7. Borucki, W.J., Koch, D., Basri, G., et al.: *Kepler* planet-detection mission: introduction and first results. Science **327**, 977–980 (2010)

8. Brown, T.M., Latham, D.W., Everett, M.E., Esquerdo, G.A.: *Kepler* input catalog: photometric calibration and stellar classification. AJ **142**, 112–130 (2011)
9. Campante, T.L., Handberg, R., Mathur, S., et al.: Asteroseismology from multi-month *Kepler* photometry: the evolved Sun-like stars KIC 10273246 and KIC 10920273. A&A **534**, 6–20 (2011)
10. Chaplin, W.J., Appourchaux, T., Elsworth, Y., et al.: The asteroseismic potential of *Kepler*: First results for solar-type stars. ApJ **713**, L169–L175 (2010)
11. Chaplin, W.J., Bedding, T.R., Bonanno, A., et al.: Evidence for the impact of stellar activity on the detectability of solar-like oscillations observed by *Kepler*. ApJ **732**, L5–L11 (2011)
12. Chaplin, W.J., Kjeldsen, H., Bedding, T.R., et al.: Predicting the detectability of oscillations in solar-type stars observed by *Kepler*. ApJ **732**, 54, 63 (2011)
13. Chaplin, W.J., Kjeldsen, H., Christensen-Dalsgaard, J., et al.: Ensemble asteroseismology of solar-type stars with the NASA *Kepler Mission*. Science **332**, 213–216 (2011)
14. Christensen-Dalsgaard, J., Kjeldsen, H., Brown, T.M., et al.: Asteroseismic investigation of known planet hosts in the *Kepler* field. ApJ **713**, L164–L168 (2010)
15. Deheuvels, S., Michel, E.: Constraints on the core mu-gradient of the solar-like star HD 49385 via low-degree mixed modes. A.N. **331**, 929–932 (2010)
16. Deheuvels, S., García, R.A., Chaplin, W.J., et al.: Seismic evidence for a rapidly rotating core in a lower-giant-branch star observed with Kepler, ApJ., in pres, ArXiv 1206.3312 (2012)
17. García, R.A., Mathur, S., Salabert, D., et al.: CoRoT reveals a magnetic activity cycle in a Sun-like star. Science **329**, 1032 (2010)
18. García, R.A., Hekker, S., Stello, D., et al.: Preparation of *Kepler* light curves for asteroseismic analyses. MNRAS **414**, L6–L10 (2011)
19. García, R.A., Ceillier, T., Campante, T., et al.: Fast rotating solar-like stars using asteroseismic datasets. ASP Conference Series in press, ArXiv 1109.6488 (2011)
20. Gilliland, R.L., Jenkins, J.M., Borucki, W.J., et al.: Initial characteristics of *Kepler* short cadence data. ApJ **713**, L160–L163 (2010)
21. Hekker, S., Broomhall, A.-M., Chaplin, W.J., et al.: The octave (Birmingham-Sheffield Hallam) automated pipeline for extracting oscillation parameters of solar-like main-sequence stars. MNRAS **402**, 2049–2059 (2010)
22. Hekker, S., Elsworth, Y., De Ridder, J., et al.: Solar-like oscillations in red giants observed with *Kepler*: comparison of global oscillation parameters from different methods. A&A **525**, 131–162 (2011)
23. Huber, D., Stello, D., Bedding, T.R., et al.: Automated extraction of oscillation parameters for *Kepler* observations of solar-type stars. Comm. Asteroseis. **160**, 74–88 (2009)
24. Huber, D., Bedding, T.R., Stello, D., et al.: Asteroseismology of red giants from the first four months of *Kepler* data: Global oscillation parameters for 800 stars. ApJ **723**, 1607–1617 (2010)
25. Kallinger, T., Mosser, B., Hekker, S., et al.: Asteroseismology of red giants from the first four months of *Kepler* data: Fundamental stellar parameters. A&A **522**, 1–14 (2010)
26. Kjeldsen, H., Bedding, T.R.: Amplitudes of stellar oscillations: the implications for asteroseismology. A&A **293**, 87–106 (1995)
27. Mathur, S., García, R.A., Régulo, C., et al.: Determining global parameters of the oscillations of solar-like stars. A&A **511**, 46–59 (2010)
28. Mathur, S., García, R.A., Catala, C., et al.: The solar-like CoRoT target HD 170987: spectroscopic and seismic observations. A&A, **518**, 53–65 (2010)
29. Mathur, S., Handberg, R., Campante, T.L., et al.: Solar-like oscillations in KIC 11395018 and KIC 11234888 from 8 months of *Kepler* data. ApJ **733**, 95–105 (2011)
30. Mathur, S., Metcalfe, T., Woitaszek, M., et al.: A uniform asteroseismic analysis of 22 solar-type stars observed by *Kepler*. ApJ **749**, 152–166 (2012)
31. Metcalfe, T.S., Monteiro, M.J.P.F.G., Thompson, M.J., et al.: A precise asteroseismic age and radius for the evolved Sun-like star KIC 11026764. ApJ **723**, 1583–1598 (2010)
32. Mosser, B., Appourchaux, T.: On detecting the large separation in the autocorrelation of stellar oscillation times series. A&A **508**, 877–887 (2009)

33. Salabert, D., García, R.A., Pallé, P.L., et al.: The onset of solar cycle 24. What global acoustic modes are telling us. A&A **504**, L1–L4 (2009)
34. Silva Aguirre, V., Chaplin, W.J., Ballot, J., et al.: Constructing a one-solar mass evolutionary sequence using asteroseismic data from *Kepler*. ApJ **740**, L2–L9 (2011)
35. Verner, G.A., Chaplin, W.J., Basu, S., et al.: Verification of the *Kepler* input catalog from Asteroseismology of solar-type stars. ApJ **738**, L28–L33 (2011)
36. Verner, G.A., Elsworth, Y., Chaplin, W.J., et al.: Global asteroseismic properties of solar-like oscillations observed by *Kepler*: a comparison of complementary analysis methods. MNRAS **415**, 3539–3551 (2011)

Part IV
Mode Identification

Identification of Pulsation Modes in Main Sequence Stars: Potentials and Limits

J. Daszyńska-Daszkiewicz and A.A. Pamyatnykh

Abstract We review the present-day methods of mode identification applied to main sequence pulsators focusing on those that make use of multicolour photometry and radial velocity data. The effects which may affect diagnostic properties of these observables are discussed. We also raise the problem of identification of high ℓ modes which can dominate oscillation spectra obtained from space-based projects.

1 Introduction

The accuracy of seismic model increases with the number of well identified pulsation frequencies. Depending on the character of modes we can probe different parts of a star and derive constraints on various parameters of model and theory. Since asteroseismology has entered the space era, the number of detected peaks in the frequency spectra has grown immensely. However, to explore these data, the unequivocal identification of mode geometry is required. In the case of main sequence pulsators this is not an easy task because their oscillation spectra do not exhibit equidistant or regular patterns. An alternative is the usage of the photometric and spectroscopic variations. In this paper, we summarize potentials and limits in using these observables for mode identification.

In Sect. 2 we recall the photometric diagrams for β Cephei, SPB (Slowly Pulsating B-type) and δ Scuti star models. We mention the most important effects which may affect diagnostic properties of such diagrams, i.e., convection,

J. Daszyńska-Daszkiewicz (✉)
Instytut Astronomiczny, Uniwersytet Wrocławski, Kopernika 11, 51-622 Wrocław, Poland
e-mail: daszynska@astro.uni.wroc.pl

A.A. Pamyatnykh
Copernicus Astronomical Center, Bartycka 18, 00-716 Warsaw, Poland
e-mail: alosza@camk.edu.pl

Daszyńska-Daszkiewicz, J. and Pamyatnykh, A.A.: *Identification of Pulsation Modes in Main Sequence Stars: Potentials and Limits.*
Astrophys Space Sci Proc. **31**, 179–184 (2013)
DOI 10.1007/978-3-642-29630-7_34,

rotation and model atmospheres. In Sect. 3 we describe a method which, beside mode identification, can yield constraints on models and theory. A prospect for identification of high ℓ modes is discussed in Sect. 4. The last section contains conclusions.

2 Mode Identification from Multicolour Photometry

Since pioneering works [1, 10, 14], the photometric observables, i.e., photometric amplitudes and phases in various passbands, have become the most often used data for mode identification in main sequence pulsators. If we ignore effects of rotation, these quantities are independent of the intrinsic mode amplitude, ε, inclination angle, i, and azimuthal order, m. In Fig. 1 we show the position of unstable modes for stellar models with masses of 12, 5 and 2 $M_\odot$ corresponding to β Cep, SPB and δ Sct stars, respectively. In the case of β Cep models we used the Strömgren uy filters and in the case of SPB and δ Sct models the Johnson BR filters. The left panels refer to all unstable modes occurring during the main sequence evolution, whereas the right panels contain unstable modes for a model with $\log T_{\rm eff} = 4.400$, 4.195 and 3.909 for the β Cep, SPB and δ Sct model, respectively. We considered modes with the degree, ℓ, up to six.

The position of pulsational modes in the photometric diagrams changes if effects of rotation are taken into account. This happens in two cases. The first case is when the frequency difference between modes is of order of the rotational frequency, and the degrees, ℓ, differ by two. Then, such modes are coupled by rotation and photometric diagrams become dependent on the inclination angle and rotational velocity [4]. The second case is when we deal with slow modes, i.e., modes with frequencies of order of the rotational frequency. Such modes are typical for the SPB pulsators. Then, the perturbation approach fails and another treatment is needed, e.g., traditional approximation. The photometric diagrams become dependent on $(i, m, V_{\rm rot})$, e.g. [8, 11]. In the case of δ Sct variables, an additional uncertainty comes from effects of the subphotospheric convection on pulsation. The use of elaborate model atmospheres (including also NLTE effects, see [9]) improves the diagnostic reliability of the photometric diagrams. As an example of the efficiency of the photometric mode identification combined with the exact fitting of observed and theoretical frequencies, we note a study of the δ Sct variable 44 Tau in which 15 oscillation frequencies have been detected. For ten of them the mode degree was uniquely determined, and the fitting of all 15 frequencies was achieved giving strong constraints on the stellar model [12, 13].

3 Including Radial Velocity Variations

The uncertainties coming from pulsation theory can be omitted, if both the mode degree, ℓ, and the nonadiabatic f-parameter are determined from observations [5, 6]. The f-parameter describes the ratio of the radiative flux perturbation to the radial

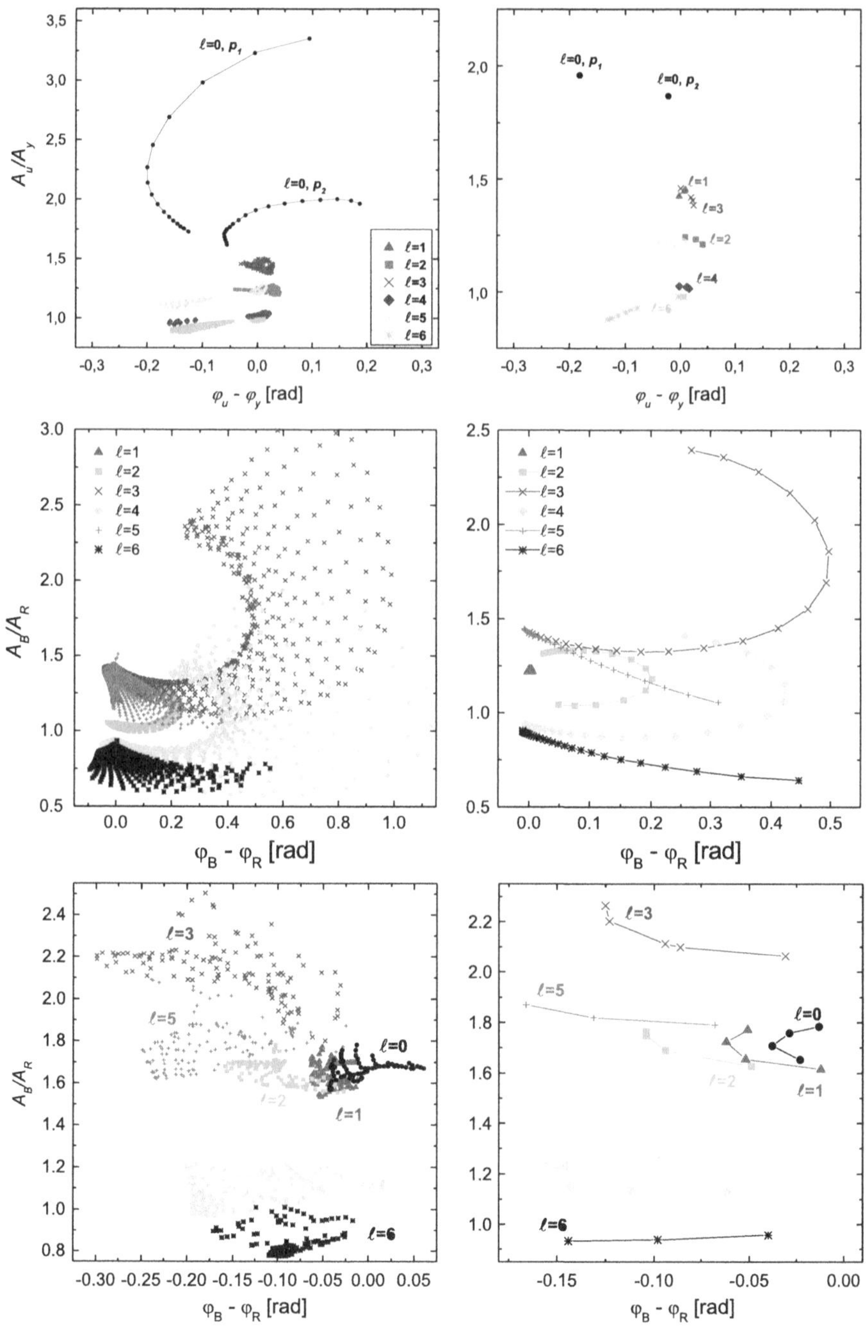

Fig. 1 The photometric diagnostic diagrams for the main sequence pulsators: β Cep stars (*top panels*), SPB stars (*middle panels*) and δ Sct stars (*bottom panels*). The *left panels* correspond to all unstable modes from ZAMS to TAMS for a stellar mass of 12, 5, and 2 $M_{\odot}$ for β Cep, SPB and δ Sct models, respectively. In the *right panels* unstable modes of particular models are shown [3]

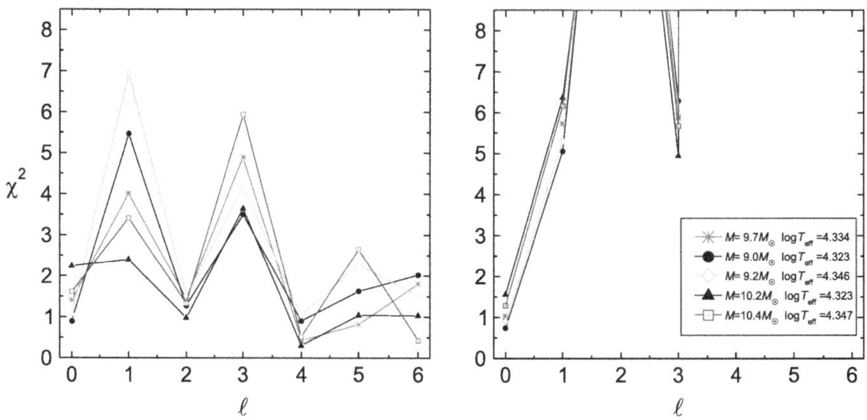

Fig. 2 The χ^2 discriminant as a function of ℓ for the β Cep star δ Ceti obtained from the fit of photometric amplitudes and phases without radial velocity data (*left panel*) and with radial velocity data (*right panel*). Different lines correspond to models from the center and edges of the error box [6]

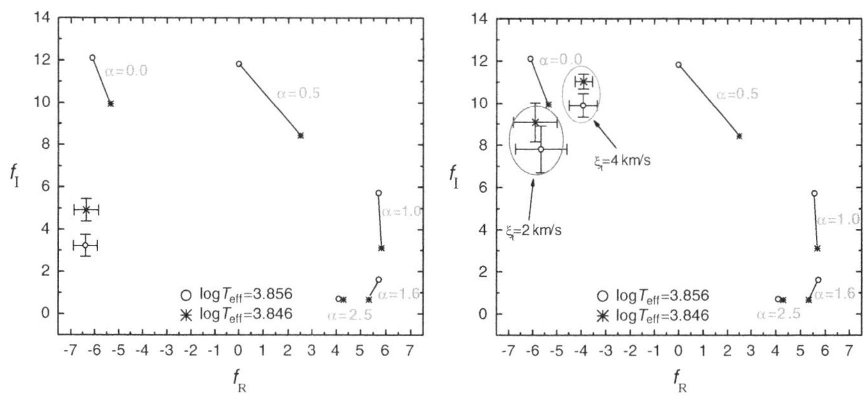

Fig. 3 Comparison of the empirical values of the f-parameter inferred from Strömgren photometry for the δ Sct star β Cassiopeiae with the theoretical ones calculated for five values of the MLT parameter, α. The empirical values of f were obtained adopting Kurucz models (*left panel*) and Vienna models (*right panel*). In the right panel the effect of the microturbulent velocity is also shown [2]

displacement at the photosphere. Figure 2 shows results for the β Cep star δ Ceti. In the case of B-type pulsators, the use of the radial velocity data is essential to get a unique identification of ℓ. Moreover, useful constraints on stellar opacities can be derived.

Modelling δ Sct stars is problematic due to the subphotospheric convection which interacts with pulsation. Mode identification using the method of simultaneous determination of ℓ and f is model independent. On the other hand, a comparison of the theoretical and empirical values of f provides useful information on the mixing-

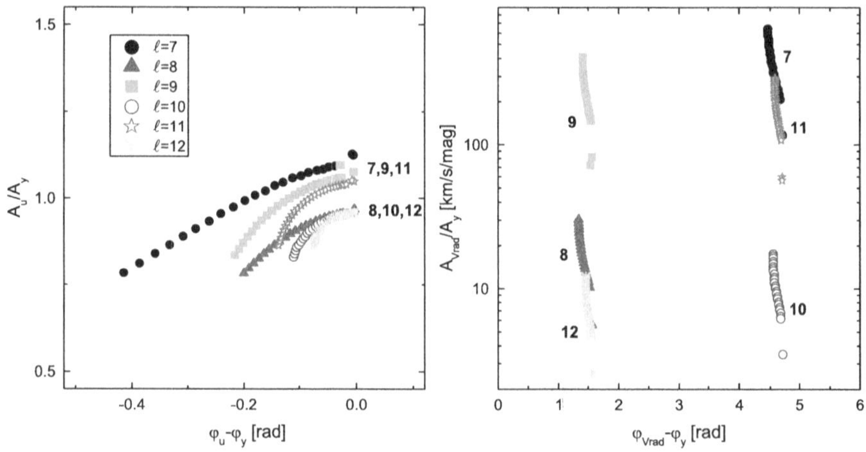

Fig. 4 Position of modes with ℓ from 7 up to 12 in the diagrams using the Strömgren uy passbands (*left panel*) and the radial velocity data and the y filter (*right panel*) for the β Cep model with $M = 8.5M_{\odot}$ and $\log T_{\rm eff} = 4.322$. Only unstable modes are shown

length parameter of convection as well as on the model atmospheres. In Fig. 3, we present such a comparison for the δ Sct star β Cassiopeiae which pulsates in the dipole mode.

4 Prospects for Extracting High ℓ Modes

The rich oscillation spectra obtained from space observations are dominated by peaks of very low amplitude. According to the simulations [7], there is a high probability that these peaks are associated to higher ℓ modes ($\ell > 6$).

Examples of diagnostic diagrams for a β Cep model and pulsational modes with ℓ from 7 up to 12 are shown in Fig. 4. The left diagram is calculated for the Strömgren uy passbands. The amplitude ratio for any two photometric passbands of the high ℓ modes is around one, because the geometrical effect, $(1-\ell)(\ell+2)$, dominates in the light variations. As we can see there is some mode grouping. This is related to the parity of degree, ℓ. The odd ℓ modes have higher values of the amplitude ratio A_u/A_y than the even ℓ modes. The right diagram shows the same modes but using radial velocity variations and the y filter. Here, the mode grouping is twofold. First, the phase differences take the values according to the sign of the disc averaging factor, b_ℓ. Modes with the phase difference value $\varphi_{Vrad} - \varphi_y$ around 1.5 rad correspond to negative values of b_ℓ, whereas those with $\varphi_{Vrad} - \varphi_y$ around 4.6 correspond to positive values of b_ℓ. Second, the values of the amplitude ratio, $A_{\rm Vrad}/A_y$, depend on the parity of ℓ. The even ℓ modes have $A_{\rm Vrad}/A_y < 30$ whereas the odd ℓ modes have $A_{\rm Vrad}/A_y > 50$. In both panels of Fig. 4, there is a separation of p and g-modes. The g-modes have negative values of $\varphi_u - \varphi_y$ and larger values of A_{Vrad}/A_y.

For much higher ℓ mode, $\ell > 20$, the amplitude ratio for any two passbands will be around one, and the phase difference around 0 for all modes. The amplitude ratios for the photometric passband and radial velocity will be spread from 0 to about 100 and the phase differences will have values of 1.5 or 4.6 rad.

5 Conclusions

In this short review, we mentioned the basic properties and uncertainties of mode identification from multicolour photometry and radial velocity data. We focused on main sequence pulsators because of their irregular oscillation spectra.

Then, we checked a prospect for identification of the high degree modes. Such modes are highly expected in the rich frequency spectra obtained from, e.g., the CoRoT, Kepler and BRITE data. Identification of the high ℓ modes only from multicolour photometry is rather hopeless. However, it turned out that it is possible to get some constraints on intermediate ℓ modes from the diagrams using amplitudes and phases of the photometric and radial velocity changes. Unfortunately, there is no prospect for extracting modes with $\ell > 30$ from these data. Here only the analysis of the line-profile variations may help.

Acknowledgements The authors acknowledge partial financial support from the Polish MNiSW grant No. N N203 379 636.

References

1. Balona, L.A., Stobie, R.S.: MNRAS **189**, 649–658 (1979)
2. Daszyńska-Daszkiewicz, J.: CoAst **150**, 32–38 (2007)
3. Daszyńska-Daszkiewicz, J.: CoAst **152**, 140–153 (2008)
4. Daszyńska-Daszkiewicz, J., Dziembowski, W.A., Pamyatnykh, A.A., Goupil, M.-J.: A& A **392**, 151–159 (2002)
5. Daszyńska-Daszkiewicz, J., Dziembowski, W.A., Pamyatnykh, A.A.: A & A**407**, 999–1006 (2003)
6. Daszyńska-Daszkiewicz, J., Dziembowski, W.A., Pamyatnykh, A.A.: A & A **441**, 641–651 (2005)
7. Daszyńska-Daszkiewicz, J., Dziembowski, W.A., Pamyatnykh, A.A.: Mem. S. A. It. **77**, 113–116 (2006)
8. Daszyńska-Daszkiewicz, J., Dziembowski, W.A., Pamyatnykh, A.A.: AA **57**, 11–32 (2007)
9. Daszyńska-Daszkiewicz, J., Szewczuk, W.: ApJ **728**, 17 (2011)
10. Dziembowski, W.: AA **27**, 203–211 (1977)
11. Dziembowski, W.A., Daszyńska-Daszkiewicz, J., Pamyatnykh, A.A.: MNRAS **374**, 249–255 (2007)
12. Lenz, P., Pamyatnykh, A.A., Breger, M., Antoci, V.: Astron. Astrophys. **478**, 855–863 (2008)
13. Lenz, P., Pamyatnykh, A.A., Zdravkov, T., Breger, M.: Astron. Astrophys. **509**, A90 (2010)
14. Watson, R.D.: Astroph. Space Sci. **140**, 255–290 (1988)

Uncovering Hidden Modes in RR Lyrae Stars

L. Molnár, Z. Kolláth, and R. Szabó

Abstract The Kepler space telescope revealed new, unexpected phenomena in RR Lyrae stars: period doubling and the possible presence of additional modes. Identifying these modes is complicated because they blend in the rich features of the Fourier-spectrum. Our hydrodynamic calculations uncovered that a "hidden" mode, the 9th overtone is involved in the period doubling phenomenon. The period of the overtone changes by up to 10% compared to the linear value, indicating a very significant nonlinear period shift caused by its resonance with the fundamental mode. The observations also revealed weak peaks that may correspond to the first or second overtones. These additional modes are often coupled with period doubling. We investigated the possibilities and occurrences of mutual resonances between the fundamental mode and multiple overtones in our models. These theoretical findings can help interpreting the origin and nature of the "hidden" modes may be found in the high quality light curves of space observatories.

1 Introduction

Before the advent of space-based, quasi-continuous observations, the light variation of RR Lyrae stars were considered to be relatively simple: they either pulsate in fundamental mode or in first overtone, or in both in the case of double-mode stars. The only complicating factor has been the Blazhko-effect, the periodic modulation of the amplitude and the phase of the pulsation. One particular model proposed that it is caused by a resonance between the fundamental mode and a large-amplitude non-radial mode [4, 10], but the resonance model was not confirmed by observations yet. Resonances that may excite overtones and cause period doubling

L. Molnár (✉) · Z. Kolláth · R. Szabó
Konkoly Observatory, Budapest, 1121, Konkoly Thege Miklós út 13-17, Hungary
e-mail: lmolnar@konkoly.hu

Molnár, L. et al.: *Uncovering Hidden Modes in RR Lyrae Stars.*
Astrophys Space Sci Proc. **31**, 185–189 (2013)
DOI 10.1007/978-3-642-29630-7_35,

were also sought but were not found between the fundamental mode and the first four overtones in a model survey [8].

2 Period Doubling

However, period doubling, the alternation of pulsation cycles with smaller and larger amplitudes was unexpectedly found in the first observations of RR Lyr itself by the *Kepler* space telescope [6]. Eventually two other stars were found where the alternations were clearly visible in the data and four were the spectrum displayed half-integer frequency peaks ($f = n/2 f_0$), the spectral features of period doubling, bringing the total count to seven [1,13]. The phenomenon was detected in modulated stars only.

Serendipitously, period doubling was also encountered in hydrodynamic models not long before the *Kepler* observations. After an extensive model survey, we were able to show that the phenomenon is caused by a 9:2 resonance between the fundamental mode and the 9th overtone [7]. The high order overtone was found to be a strange mode. The resonance displays surprising strength: while the period of the pulsation cycle is shifted by less than 0.5% compared to the linear period value of the fundamental mode, the period of the 9th overtone can be changed by as much as 5–10%.

3 Additional Modes

Significant peaks besides the pulsation frequency, it's harmonics and the modulation side lobes were detected in RR Lyrae Fourier spectra both with the CoRoT [3,5,12] and *Kepler* space telescopes [1]. They appear close to the period ratios of the first or second overtones, but the amplitude difference compared to the fundamental mode is so large that such stars cannot be considered as classical double mode pulsators. Out of the six stars with similar additional peaks, four show both the Blazhko-effect and period doubling too. Two stars however (KIC 7021124 and 9508655) display only marginal signs of modulation with side lobe amplitudes below 1.5 and 0.5 mmag respectively, that can be attributed to the residual amplitude differences between the different quarters of the data as well [1,9].

An interesting case is V354 Lyr (KIC 6183128), where the $nf_0 + f_2$ combination peaks do not decrease monotonically, as expected, suggesting some resonance with an undetected mode (Fig. 1). The increase occurs in the frequency region of the 9th overtone, the strange mode. Our current hypothesis is a three-mode resonance between the fundamental mode and the two overtones, in the from of $3f_0 + f_2 = f_9$. The position of the linear three-mode resonance region in our diagnostic diagram (T_{eff} vs. $150\,M - L$, see [7]) is displayed in Fig. 2.

The period ratios of the fundamental and additional modes can be used to determine the physical characteristics of these stars. We could not find nonlinear

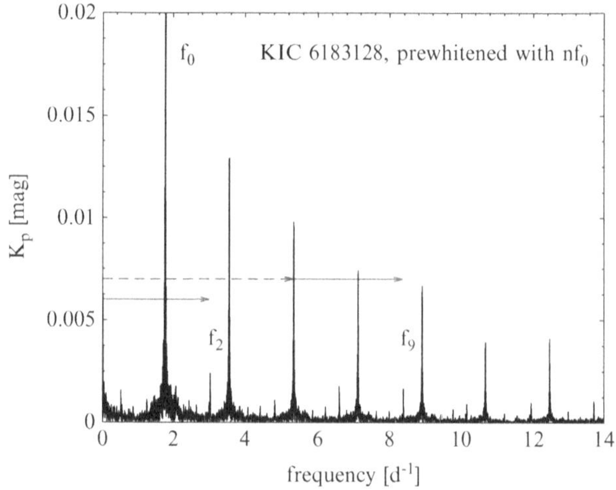

Fig. 1 Fourier-spectrum of *Kepler* Q1-Q2 data of V354 Lyr (KIC 6183128), pre-whitened with the pulsation frequency and it's harmonics. Note the higher amplitude $nf_0 + f_2$ peaks at 6.6 and 8.4 c/d. The latter may indicate the presence of a resonant f_9 mode. *Solid and dashed arrows* visualise the possible $3f_0 + f_2 = f_9$ resonance

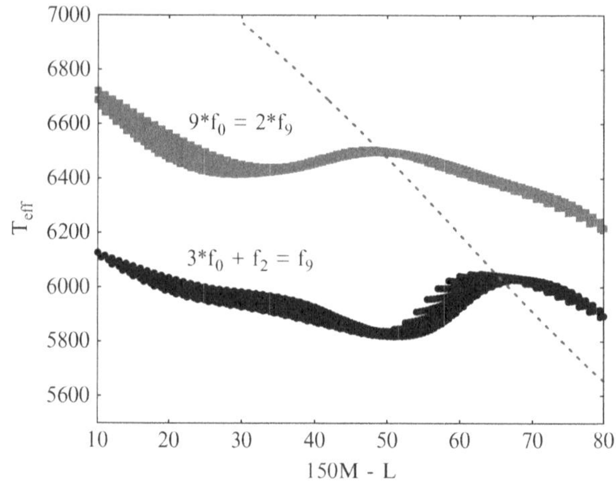

Fig. 2 The diagnostic diagram with two resonance regions included. Areas covered with rectangles and points represent linear models where $9f_0 = 2f_9$ and $3f_0 + f_2 = f_9$, respectively. The former is responsible for the period doubling. The *dashed line* shows the linear models fitted to the f_0 and f_2 frequency values of V354 Lyr (KIC 6183128)

models with a three-mode resonance and excited additional modes yet, however the strong period shifts we found in the 9:2 resonance case indicate that resonant nonlinear models may be located at a considerably different parameter range compared to the intersection of the linear results, displayed in Fig. 2.

4 Future Prospects

Radial resonances gained more interest after the emergence of a new Blazhko-model. The 9:2 resonance, incorporated in amplitude equations, produced not only period doubling but also amplitude modulation [2]. It is unclear however, how far are the modulated AE solutions from the hydrodynamic models of RR Lyrae stars. Finding more resonances and fitting the corresponding coefficients of AEs from hydrodynamic calculations might help such investigations.

Kepler revealed a whole variety of additional frequency peaks in RR Lyrae stars around and below the milli magnitude level. Half of the modulated stars in the *Kepler* RRab sample display either period doubling or other signs or both. Additional peaks are not limited to the first or second overtone candidates but are found at different values as well, suggesting the presence of additional resonances and the possibility of nonradial modes too.

From the theoretical side, we also detected peculiar hydrodynamic solutions with irregularly varying amplitudes that were found to be chaotic solutions [11]. The *Kepler* long cadence observations hinted some irregularity in period doubling that might be the manifestation of this chaotic behaviour but short cadence observations and careful analysis will be needed to settle this issue.

Acknowledgements This work was supported by the Hungarian OTKA grants K83790 and MB08C 81013. RSz acknowledges the Bolyai János Scholarship of the Hungarian Academy of Sciences. The research leading to these results has received funding from the European Community's Seventh Framework Programme (FP7/2007–2013) under grant agreement no. 269194. Funding for the *Kepler* mission is provided by the NASA Science Mission directorate.

References

1. Benkő, J.M., Kolenberg, K., Szabó, R., et al.: Flavours of variability: 29 RR Lyrae stars observed with Kepler. MNRAS **409**, 1585–1593 (2010)
2. Buchler, J.R., Kolláth, Z.: On the Blazhko effect in RR Lyrae stars. ApJ **731**, 24–28 (2011)
3. Chadid, M., Benkő, J.M., Szabó, R., et al.: First CoRoT light curves of RR Lyrae stars. Complex multiplet structure and non-radial pulsation detections in V1127 Aquilae. A&A **510**, 39–49 (2010)
4. Dziembowski, W., Cassisi, S.: Nonradial modes in RR LYR stars. AcA **49**, 371–382 (1999)
5. Guggenberger, E., Kolenberg, K., Chapellier, E., et al: The CoRoT star 105288363: strong cycle-to-cycle changes of the Blazhko modulation. MNRAS **415**, 1577–1589 (2011)
6. Kolenberg, K., Szabó, R., Kurtz, D.W., et al.: First Kepler results on RR Lyrae stars. ApJL **713**, L198–L203 (2010)
7. Kolláth, Z., Molnár, L., Szabó, R.: Period-doubling bifurcation and high-order resonances in RR Lyrae hydrodynamical models. MNRAS **414**, 1111–1118 (2011)
8. Moskalik, P., Buchler, J.R.: Resonances and period doubling in the pulsations of stellar models. ApJ **355**, 590–601 (1990)
9. Nemec, J.M., Smolec, R., Benkő, J.M., et al.: Fourier analysis of non-Blazhko ab-type RR Lyrae stars observed with the Kepler space telescope. MNRAS *417*, 1022–1053 (2011)

10. Nowakowski, R.M., Dziembowski, W.A.: Resonant excitation of nonradial modes in RR Lyr stars. AcA **51**, 5–47 (2001)
11. Plachy, E., Kolláth, Z., Molnár, L.: Detection of chaos in RR Lyrae hydrodynamical models. These proceedings (2011)
12. Poretti, E., Paparó, M., Deleuil, M., et al.: CoRoT light curves of RR Lyrae stars. CoRoT 101128793: long-term changes in the Blazhko effect and excitation of additional modes. A&A **520**, 108–114 (2010)
13. Szabó, R., Kolláth, Z., Molnár, L., et al.: Does Kepler unveil the mystery of the Blazhko effect? First detection of period doubling in Kepler Blazhko RR Lyrae stars. MNRAS **409**, 1244–1252 (2010)

Complex Asteroseismology of the Slowly Pulsating B-Type Star HD74560

P. Walczak, W. Szewczuk, and J. Daszyńska-Daszkiewicz

Abstract We present the results of complex seismic modelling of the Slowly Pulsating B-type star HD74560. The star pulsates in five frequencies detected in photometric observations. For all these frequencies, we identify the mode degree, ℓ. For two of them, found also in spectroscopic data, we are able to derived the empirical values of the complex nonadiabatic parameter f. We test effects of chemical composition and opacity data. Our results show that the properties of seismic models of SPB stars differ significantly from those of the more massive β Cephei stars.

1 Introduction

The Slowly Pulsating B-type stars (SPB) are very interesting targets for asteroseismic analysis. They pulsate in high-order g-modes, which penetrate the deep interior of the star. Because of it, these types of modes are extremely useful for studying the overshooting efficiency. This effect is poorly understood and analysing pulsations of the SPB stars is a unique opportunity to get constraints on its range.

HD74560 is a bright object with the parallax of $\pi = 6.73(17)$ mas. [7] discovered line profile variations with two frequencies and classified the star as a SPB variable. He also found two candidates for a third frequency but it was impossible to determine the physical one owing to the aliasing. At least five frequencies in the Geneva photometry were found by [4]. Only the dominant frequency, ν_1, agrees with the results of [7]. Also, two of their five frequencies in the radial velocity variations [4].

P. Walczak (✉) · W. Szewczuk · J. Daszyńska-Daszkiewicz
Instytut Astronomiczny, Uniwersytet Wrocławski, Kopernika 11, 51-622 Wrocław, Poland
e-mail: szewczuk@astro.uni.wroc.pl; walczak@astro.uni.wroc.pl; daszynska@astro.uni.wroc.pl

Walczak, P. et al.: *Complex Asteroseismology of the Slowly Pulsating B-Type Star HD74560*. Astrophys Space Sci Proc. **31**, 191–195 (2013)
DOI 10.1007/978-3-642-29630-7_36,

In addition to the fitting frequencies our seismic analysis makes use of the complex nonadiabatic f-parameter, defined as the ratio of the bolometric flux changes to the radial displacement (see i.e. [1, 3]). Fitting frequencies and the corresponding values of f is termed *complex asteroseismology* and it can provide stronger constraints on the model parameters and theory.

The paper is organized as follows. Results of mode identification are shown in Sect. 2. In Sect. 3 we present our attempt towards complex seismic modeling of the SPB star HD74560. Conclusions end the paper.

2 Mode Identification

The five frequencies of HD74560 are shown in the first column of Table 1. These values were taken from [4]. The frequencies range from about 0.4 up to 0.6 c/d which is typical for the SPB stars.

In order to identify the degree, ℓ, of pulsational modes we incorporated the theoretical and empirical values of the f-parameter. In the first case we compared the theoretical amplitude ratios and phase differences in different passbands with observational counterparts. The appropriate theoretical values of f result from linear nonadiabatic calculations of stellar pulsations. We used the seven Geneva passbands and the non-LTE models of stellar atmospheres by [6]. The identified mode degrees are presented in the third column of Table 1. As we can see, the most probable identification for all frequencies are the dipole modes, $\ell = 1$. The ν_2 and ν_3 frequencies can be also the quadruple modes, $\ell = 2$.

In the second method, we derived the mode degrees together with the empirical values of the f-parameter. We have to stress that in the case of the B-type pulsators, we are able to extract the empirical f only for modes which are seen both in photometry and spectroscopy [2]. Therefore, by means of this method we identified the mode degrees only for two frequencies, ν_1 and ν_2. The results, given in the last column of Table 1, are in agreement with mode degrees derived with the previous method. The only difference is for ν_1; it can be the quadruple mode as well.

Summarizing, all pulsational frequencies of HD74560 can be the dipole modes, while ν_2 and ν_3 can also be the quadruple modes.

3 Complex Asteroseismology

In the previous section we determined the mode degrees, ℓ, of the observed frequencies of HD74560. In the next step, we constructed models reproducing the frequencies. Firstly, we chose two modes, ν_1 and ν_4. The ν_1 frequency is the strongest, well identified mode, and the ν_4 frequency was chosen because it has the highest value. Relatively high values of frequencies facilitate modelling. Apart

Table 1 Pulsational frequencies of HD74560 with corresponding values of the Geneva V amplitudes and the most probable mode identification derived with the theoretical and empirical values of the f-parameter

Frequency (c/d)	The Geneva V amplitude (mmag)	Mode degree, ℓ, theoretical f-parameter	Mode degree, ℓ, empirical f-parameter
$\nu_1 = 0.64472$	14.4(4)	1	1,2
$\nu_2 = 0.39578$	4.3(4)	1,2	1,2
$\nu_3 = 0.44763$	3.2(4)	1,2	–
$\nu_4 = 0.82281$	3.7(4)	1	–
$\nu_5 = 0.63567$	3.0(4)	1	–

from that, ν_2, ν_3, ν_4 and ν_5 have rather similar amplitudes (see the second column of Table 1).

In order to construct seismic models we have to know the azimuthal number, m, and radial order, n, of our modes. We assumed that pulsational modes of HD74560 are axisymmetric. The star is rather a slow rotator, ($V \sin i = 35$ km/s according to [5]), and the results should not differ significantly from those presented here.

The frequency spectrum of high-order g-modes is very dense and identification of the radial order is difficult, but it turned out that fitting two frequencies constrains n quite strongly. We were able to find seismic models only for two combinations of the radial orders. The first possibility is: ν_1 is the $\ell = 1$, g_{22} mode and ν_4 is the $\ell = 1$ g_{17} mode. The second one is: ν_1 is the $\ell = 1$ g_{17} mode and ν_4 is the $\ell = 1$ g_{13} mode.

In Fig. 1, we show the position of seismic models of HD74560 on the HR diagram for several values of the metallicity parameter, Z. These models fit the ν_1 and ν_4 frequencies assuming the first combination of the radial orders. We stress, that for a given Z there are no seismic models between points shown in Fig. 1. The error box was taken from [4] and Pamyatnykh (private communication). The left and right panels show the results for the OP and OPAL opacity tables, respectively. All models are computed without any overshooting of the convective core. The hatched and grey areas indicate the instability regions of the ν_1 and ν_4 modes, respectively. The areas labeled as "f-parameter" indicate models fitting the empirical value of f for the ν_1 mode.

As we can see, the effective temperature of seismic models strongly depends on metallicity. Both with the OP and OPAL opacities, we need quite low value of Z to catch models inside the observational error box ($Z < 0.007$). The instability regions computed with different opacities differ significantly. In the case of the OPAL tables we need much higher value of Z to excite the ν_1 and ν_4 modes ($Z > 0.007$). With the OP data, models with metallicity of $Z = 0.006$ are already unstable.

Relying on the f parameter, it is difficult to judge which opacity data are more suitable. Although with the OPAL tables, the f-parameter area is larger, most of the models fitting f are located outside of the error box.

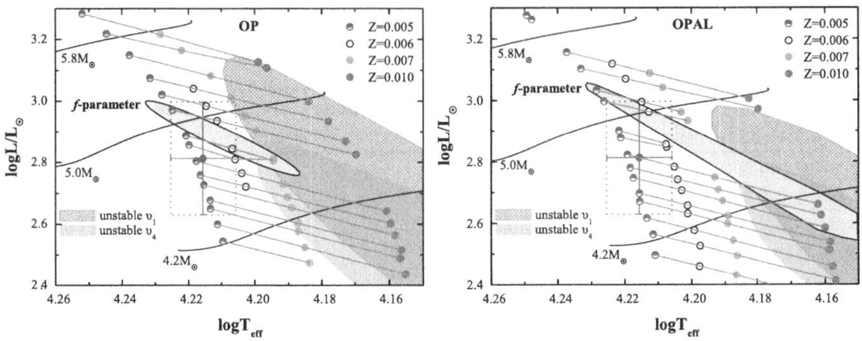

Fig. 1 Seismic models of the SPB star HD74560 fitting two frequencies, ν_1 and ν_4, on the HR diagram. Results for four values of metallicity, Z, are shown. The *left and right panels* show models obtained with the OP and OPAL data, respectively. We marked also regions where the ν_1 and ν_4 modes are pulsational unstable

We also performed seismic analysis assuming $\alpha_{\rm ov} = 0.1$, 0.2 and 0.4. In these cases, we did not find any model fitting the empirical f-parameter within 1σ errors. Thus, the f-parameter can be used to constrain the overshooting range.

Interesting results were obtained with the second combination of radial orders: ν_1 is the $\ell = 1$ g_{17} mode and ν_4 is the $\ell = 1$, g_{13} mode. In this case, seismic models located inside the observational error box have much higher metallicity, $Z \approx 0.014$. As a consequence, all models fitting the observational effective temperature and luminosity have unstable modes ν_1 and ν_4 . Moreover, inside of the error box there is at least one model fitting the empirical value of the f-parameter for the ν_1 mode.

4 Conclusions

Seismic modelling of the SPB stars is quite challenging. There are seismic models inside the error box computed with the OP and OPAL opacities and with different combinations of radial orders. Some of these models reproduce also the empirical value of the f-parameter, but to get stronger constraints on stellar parameters, we need to fit more than two frequencies.

For the time being, we can conclude that there are some preferences towards the OP data owing to the larger instability areas for ν_1 and ν_4. All seismic models preferred small overshooting parameter, $\alpha_{\rm ov} \lesssim 0.1$. Interesting constraints were derived for metallicity. The values of Z can be either very small, ~0.005–0.007 (assuming the first combination of the radial orders), or quite high ~0.013–0.015 (for the second combination of n). The other values of Z are excluded.

Acknowledgements The authors acknowledge partial financial support from the Polish MNiSW grant No. N N203 379 636.

References

1. Daszyńska-Daszkiewicz, J., Dziembowski, W.A., Pamyatnykh, A.A.: A&A **407**, 999 (2003)
2. Daszyńska-Daszkiewicz, J., Dziembowski, W.A., Pamyatnykh, A.A.: A&A **441**, 641 (2005)
3. Daszyńska-Daszkiewicz, J., Walczak, P.: MNRAS **398**, 1961 (2009)
4. De Cat, P., Aerts, C.: A&A **393**, 965 (2002)
5. Głebocki, R., Stawikowski, A.: AcA **50**, 509 (2000)
6. Lanz, T., Hubeny, I.: ApJS **169**, 83 (2007)
7. Waelkens, C.: A&A **246**, 453 (1991)

Rotation Effects as Possible Responsible for the Surface Effects on the Sun and Solar-Type Stars

J.C. Suárez, M.J. Goupil, D.R. Reese, R. Samadi, F. Lignières, M. Rieutord, and J. Lochard

Abstract One of the main sources of uncertainty in the asteroseismic models of the Sun and solar-like stars is the poor match between predicted oscillation frequencies and observed ones in the very high frequency domain. Today, effects of turbulence, diffusion, etc., i.e., the so-called "surface effects" are signaled as possible responsible for such a discrepancy. We show that the effect of the stellar deformation due to rotation is of the same order or even larger than these effects. We show that rotation effects, are important for the asteroseismic analysis of the Sun and solar-like stars and they cannot be neglected when modeling such stars.

1 Introduction

Today, stellar physics, and in particular the asteroseismology, witnesses a significant boost [30] thanks to the space, where satellites like *MOST* [41], *CoRoT* [1], and *Kepler* [13], are providing asteroseismic data with unprecedented accuracy, see e.g. [2, 3, 5, 8, 9, 12, 21] to name a few.

The effect of rotation is one of the main source of uncertainty in asteroseismology. It affects the form of the resonant cavity and plays an important role in almost

J.C. Suárez (✉)
IAA-CSIC, Glorieta de la Astronomía S/N, Granada, Spain
e-mail: jcsuarez@iaa.es

M.J. Goupil · D.R. Reese · R. Samadi · J. Lochard
Observatoire de Paris, LESIA, CNRS UMR 8109, 92195 Meudon, France
e-mail: mariejo.goupil@obspm.fr; daniel.reese@obspmfr; reza.samadi@obspm.fr; jeremie.lochard@gmail.com

F. Lignières · M. Rieutord
Université de Toulouse, UPS, Laboratoire d'Astrophysique de Toulouse-Tarbe (LATT), 31400 Toulouse, France
e-mail: francois.lignieres@ast.obs-mip.fr; michel.rieutord@ast.obs-mip.fr

Suárez, J.C. et al.: *Rotation Effects as Possible Responsible for the Surface Effects on the Sun and Solar-Type Stars*. Astrophys Space Sci Proc. **31**, 197–200 (2013)
DOI 10.1007/978-3-642-29630-7_37,

all physical process occurring during stellar evolution [15]. For slow-to-moderately rotating stars, like solar-like stars, the oscillation computation, it has been shown that rotation effects cannot be neglected for asteroseismic studies [4,6,7,14,22,24–26, 32, 33, 38]. Analysis of low-mass models for different turbulent diffusion coefficients [17] revealed that rotation may increase the horizontal transport and the rotational effects upon frequencies. Moreover, rotationally-induced mixing in solar-type stars is not strong enough to impose a rigid rotation, as opposed to the solar case [10].

Excited modes in solar-like stars are high-frequency p modes with small inertia. These propagate through the outer layers of the star, hence they frequencies are more sensitive to changes in the surface physical properties, where the centrifugal force becomes more efficient. The present work analyzes the effects of rotation in this high-frequency domain, and its impact on the different seismic diagnostics techniques based on the asymptotic properties of the oscillations. The main results have been published in the paper [36] (hereafter S10).

2 Peturbative vs. Non-perturbative Approaches

This work has been performed using 1D asteroseismic models, computed using the evolutionary code CESAM [19], and the adiabatic oscillation code FILOU [29, 34, 35], following the prescriptions by the ESTA/CoRoT team [20]. The perturbative approach is only valid for a certain range of star distortion and frequency domain as explained in [16,23] from which a complete calculation (non-perturbative approach) is necessary. The first semi-empirical determination of such a limit was obtained for the δ Scuti star Altair [31], but no equivalent had been done so far for solar-type stars. We show that the P approach is valid up to $v_{\rm rot} = 40$–$50\,{\rm km\,s^{-1}}$ (more details in S10).

3 Effect of Rotation on Diagnostics of Echelle Diagrams

Figure 1 shows echelle (details in S10) diagrams calculated for a realistic model of a solar-like star (see Table 1 in S10), assuming a uniform rotation of 32 kms. Ridges are altered significantly towards the highest frequency domain. For rotational velocities of about $16\,{\rm km\,s^{-1}}$ and higher, diagnostics on large spacings and on modal identification through echelle diagrams can be altered by the presence of the $m \neq 0$ components of the rotationally split modes. These effects are detectable in the observed frequency range (from $\nu_{\rm max}$ to higher frequencies). We found that the larger the rotational velocity of the star is, the more chances the observed ridges in echelle diagrams correspond to $\ell = 0$ and/or $\ell = |m|$ modes. This is in agreement with the the empirical correlation found by [37] between the amplitudes of p modes with the angle of inclination i (details in S10). The effect of rotation on

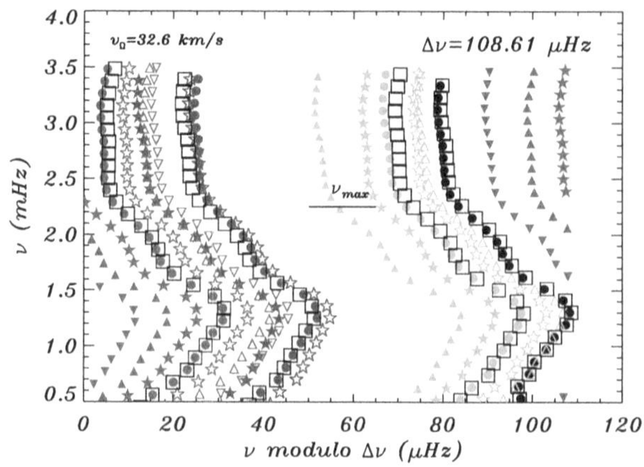

Fig. 1 Echelle diagrams for the uniformly-rotating models. *Open squares* correspond to ridges of the non-rotating stellar reference model. *Black and green symbols* represent $\ell = 0$ and $\ell = 2$ modes, respectively. *Blue and red symbols* represent $\ell = 1$ and $\ell = 3$ modes, respectively. *Filled circles* represent both $\ell = 0$ and $m = 0$ modes. For the rest of modes, *filled and empty symbols* represent the $-m$ and $+m$ frequencies, respectively. *Stars, triangles, and inverted triangles* represent modes with $|m| = 1, 2$, and 3, respectively. The *small horizontal line* shows the location of the $\nu_{\max}$ frequency. Figure taken from S10

the oscillations may go from a few muHz up to 20–30 μHz for the largest rotational velocity considered (around 50 km s^{-1}). This is of the same order or even larger than the so-called *surface effects*. These later are, for instance, the turbulence [28] (up to 5 μHz, from $\nu_{\max}$ to higher frequencies). Moreover microscopic diffusion might also contribute to such surface effects [39, 40] which are of the order of a few μHz.

In addition, other wisely used diagnostic tools based on frequency combinations in asymptotic regime are also altered, so their use, and more importantly the physical conclusions obtained with them must be considered with care (see Sect. 2 of S10).

4 Conclusions

We have shown that rotation must be considered as one of the so-called surface effects of the Sun and solar-like stars. For these stars, rotation effects are of the order or even larger than other physical effects considered to explain the so-called surface effects. As a consequence, this indicates that the Sun asteroseismic standard model should be revised.

This result might also be important for asteroseismic studies aiming at characterizing with high accuracy the exo-planetary systems around solar-like stars (see [11, 18, 27], as examples of studies based on Kepler data).

Acknowledgements JCS acknowledges support from the "Instituto de Astrofísica de Andalucía (CSIC)" by an "Excellence Project post-doctoral fellowship" financed by the Spanish "Conjerería de Innovación, Ciencia y Empresa de la Junta de Andalucía" under project "FQM4156-2008". JCS also acknowledges support by the Spanish "Plan Nacional del Espacio" under project ESP2007-65480-C02-01. DRR acknowledges support from the CNES through a post-doctoral fellowship.

References

1. Baglin, A.: Adv. Space Res. **31**, 345 (2003)
2. Balona, L.A., Pigulski, A., Cat, P.D., et al.: MNRAS **413**, 2403 (2011)
3. Breger, M., Balona, L., Lenz, P., et al.: MNRAS **414**, 1721 (2011)
4. Briquet, M., Uytterhoeven, K., Morel, T., et al.: A&A **506**, 269 (2009)
5. Bruntt, H., Stello, D., Suárez, J.C., et al.: MNRAS **378**, 1371 (2007)
6. Casas, R., Moya, A., Suárez, J.C., et al.: ApJ **697**, 522 (2009)
7. Casas, R., Suárez, J.C., Moya, A., Garrido, R.: A&A **455**, 1019 (2006)
8. Catanzaro, G., Ripepi, V., Bernabei, S., et al.: MNRAS **411**, 1167 (2011)
9. Chaplin, W.J., Appourchaux, T., Elsworth, Y., et al.: ApJ **713**, L169 (2010a)
10. Eggenberger, P., Carrier, F.: A&A **449**, 293 (2006)
11. García, R.A., Salabert, D., Ballot, J., et al.: J. Phys. Conf. Ser. **271**, 012046 (2011)
12. García Hernández, A., Moya, A., Michel, E., et al.: A&A **506**, 79 (2009)
13. Gilliland, R.L., Brown, T.M., Christensen-Dalsgaard, J., et al.: PASP**122**, 131 (2010)
14. Goupil, M.J.: In: Rozelot, J.-P., Neiner, C. (eds.) Lecture Notes in Physics, vol. 765, pp. 45–99. Springer, Berlin (2009)
15. Goupil, M.-J., Dupret, M.A., Samadi, R., et al.: J. Astrophys. Astr. **26**, 249 (2005)
16. Lignières, F., Rieutord, M., Reese, D.: A&A **455**, 607 (2006)
17. Mathis, S., Palacios, A., Zahn, J.-P.: A&A **462**, 1063 (2007)
18. Metcalfe, T.S., Monteiro, M.J.P.F.G., Thompson, M.J., et al.: ApJ **723**, 1583 (2010)
19. Morel, P.: A&AS **124**, 597 (1997)
20. Moya, A., Christensen-Dalsgaard, J., Charpinet, S., et al.: Ap&SS **316**, 231 (2008)
21. Poretti, E., Rainer, M., Weiss, W.W., et al.: A&A **528**, A147+ (2011)
22. Poretti, E., Suárez, J.C., Niarchos, P.G., et al.: A&A **440**, 1097 (2005)
23. Reese, D., Lignières, F., Rieutord, M.: A&A **455**, 621 (2006)
24. Rodríguez, E., Amado, P.J., Suárez, J.C., et al.: A&A **450**, 715 (2006a)
25. Rodríguez, E., Costa, V., Zhou, A.-Y., et al.: A&A **456**, 261 (2006b)
26. Soufi, F., Goupil, M.J., Dziembowski, W.A.: A&A **334**, 911 (1998)
27. Stello, D., Chaplin, W.J., Bruntt, H., et al.: ApJ **700**, 1589 (2009)
28. Straka, C.W., Demarque, P., Guenther, D.B., Li, L., Robinson, F.J.: ApJ **636**, 1078 (2006)
29. Suárez, J.C.: Ph.D. Thesis, ISBN 84-689-3851-3, ID 02/PA07/7178 (2002)
30. Suárez, J.C.: Lecture Notes Essays Astrophys. **4**, 33 (2010)
31. Suárez, J.C., Bruntt, H., Buzasi, D.: A&A **438**, 633 (2005)
32. Suárez, J.C., Garrido, R., Goupil, M.J.: A&A **447**, 649 (2006a)
33. Suárez, J.C., Garrido, R., Moya, A.: A&A **474**, 961 (2007)
34. Suárez, J.C., Goupil, M.J.: Ap&SS **316**, 155 (2008)
35. Suárez, J.C., Goupil, M.J., Morel, P.: A&A **449**, 673 (2006b)
36. Suárez, J.C., Goupil, M.J., Reese, D.R., et al.: ApJ **721**, 537 (2010)
37. Suárez, J.-C., Michel, E., Pérez Hernández, F., et al.: A&A **390**, 523 (2002)
38. Suárez, J.C., Moya, A., Amado, P.J., et al.: ApJ **690**, 1401 (2009)
39. Théado, S., Vauclair, S., Castro, M., Charpinet, S., Dolez, N.: A&A **437**, 553 (2005)
40. Thoul, A., Montalbán, J.: Microscopic diffusion in stellar plasmas. In: Straka, C.W., Lebreton, Y., Monteiro, M.J.P.F.G. (eds.), vol. 26, pp. 25–36. EAS Publication Series (2007)
41. Walker, G., Matthews, J., Kuschnig, R., et al.: PASP **115**, 1023 (2003)

Theoretical Approach to Mode Identification

Joyce Ann Guzik

Abstract Mode identification in pulsating stars is challenging because the modes that are predicted to be excited and visible are not all observed, and because sometimes modes that are not expected are observed. In principle, finding rotationally split multiplets can assist mode identification, but often not all of the components are observed, and rapid and differential rotation complicates the interpretation. Other challenges include distinguishing pulsations from star spots, identifying frequencies that are linear combinations of other (perhaps invisible) intrinsic modes, mode coupling, and variable mode amplitudes and frequencies. For brighter stars and modes with high signal-to-noise, spectroscopic and photometric techniques have had some success in separating $l = 0$, 1 and 2 modes and in identifying the azimuthal orders. The nearly equal frequency (period) spacings for high order p- (g-) mode pulsators expected from asymptotic theory can guide mode identifications. We review theoretical expectations for pulsation mode driving and damping, focusing on main-sequence variables, and compare with observational examples. Insights into mode selection and amplitudes may be possible by examining the energy partition between various processes in these stars and their contributions to driving and damping of the oscillation modes. Future progress will require two- and three-dimensional stellar models and nonadiabatic, nonlinear, and nonradial pulsation modeling.

1 Expectations for Main Sequence Variables

We review the modes expected for normal main-sequence variables: Solar-like, γ Doradus, δ Scuti, Slowly Pulsating B, and β Cep stars. We focus on single stars, and neglect tidal forcing (e.g. as in KOI-54), roAp stars (e.g. HR 1217) where

J.A. Guzik (✉)
Los Alamos National Laboratory, XTD-2, MS T-06, Los Alamos NM 87545, USA
e-mail: joy@lanl.gov

Guzik, J.A.: *Theoretical Approach to Mode Identification.*
Astrophys Space Sci Proc. **31**, 201–208 (2013)
DOI 10.1007/978-3-642-29630-7_38,

magnetic fields are important, and the chemically peculiar Am and λ Boo stars. We discuss complications for mode identification, the lack of agreement between observation and expectation, and what we do in practice to identify modes. We will not review photometric and spectroscopic identification methods (but see [1] for a review of these).

Mode identification for the Sun and solar-like stars is probably the most tractable and straightforward. These stars are slowly rotating, and high-order p modes are excited stochastically and detectable in photometry and radial velocity observations. Usually the entire sequence of consecutive radial orders is observed within the envelope of power. Asymptotic theory predicts approximately equal frequency spacing for consecutive radial orders. Small and large separations can be identified. Many examples of solar-like power spectra are available now, with a progression in peak power to lower frequency with lower mean density [1, 10, 18]. Echelle diagrams, plotting frequency modulo the large frequency separation, show characteristic patterns of $l = 0, 2$ and $l = 1, 3$ modes aligned in ridges. Being certain of which ridge set corresponds to $l = 1, 3$ and which to $l = 0, 2$ and also of the radial order of the modes can be difficult, and other constraints are needed (e.g. stellar luminosity, or comparing the shape of ridges in the Echelle diagram with that expected from trial models).

The γ Dor stars with masses $\sim$1.4–1.7 $M_\odot$ pulsate in high-order g modes, driven by the convective blocking mechanism near the envelope convection-zone base at $\sim$200–500,000 K. Asymptotic theory for high-order g modes predicts equal *period* spacings. Line-profile variations can be used to identify modes with higher l and m not visible in full-disk photometry.

We calculated the equally spaced periods for the $l = 1$ and 2 g modes predicted for a 1.48 $M_\odot$, $Z = 0.02$ γ Dor model with age $\sim$1 Gyr that has exhausted about half its core hydrogen, and temperature of $\sim$350,000 K at the base of envelope convection-zone. Figure 1 (left) plots the model's *frequencies* (which are not equally spaced) vs. linear growth rate per period. If these predictions are compared to a not-atypical *Kepler* γ Dor star KIC 08222685 (Fig. 1, right), there is some similarity to the double-peaked pattern. While it is suggestive, there is no reason to expect that the modes with high predicted linear growth rates also will have high amplitudes. Qualitative agreement between predicted linear growth rates and observed amplitude does not hold, for example, for the δ Scuti star FG Vir. It would be useful to make Echelle diagrams of periods modulo the *period* spacing for γ Dor stars to compare with a calculated Echelle diagram.

At masses $\sim$1.8–2 $M_\odot$ the δ Scuti stars pulsate via the κ-effect driving from the second ionization of helium at $\sim$50,000 K. Low-order p modes are predicted, and modes of mixed character, with p-type nodes in the envelope and g-type nodes in the interior [1, 18]. Avoided crossings (seen in frequency vs. age diagrams) are present because of the development of a region where g modes can propagate when composition gradients develop outside the convective core. Therefore, finding such signatures in the frequency spectrum is very useful to diagnose convective core sizes and stellar age.

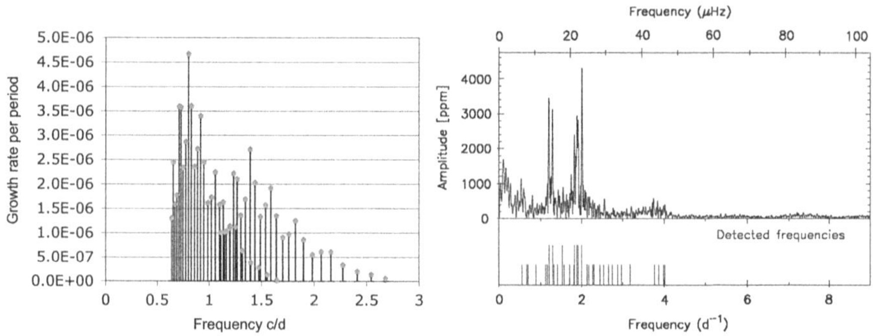

Fig. 1 Predicted $l = 1$ and $l = 2$ g-mode spectrum for a 1.48 $M_{\odot}$ model vs. linear growth rate per period (*left*); observed frequency spectrum vs. amplitude for KIC08222685 [44] (*right*). A double peaked spectrum as in KIC08222685 is often seen in *Kepler* γ Dor stars. While it is suggestive, there is no reason to assume that the modes with the highest predicted growth rates also will have high amplitudes. Qualitative agreement between predicted linear growth rates and observed amplitude does not hold, for example, for the δ Scuti star FG Vir

δ Sct stars (and higher-mass B stars) can have relatively rapid rotation of ~50–100 km/s, and rotational frequency splitting can aid or complicate the mode identification. Perturbative treatments expanding in rotation frequency give, to first order, equal splitting of a mode into $2l+1$ equally spaced multiplets (Coriolis term); to second order, these frequencies, including the $m = 0$ mode, are further shifted (centrifugal term) [28]. Ideally, one could use this rotational splitting pattern to determine the stellar rotation rate and to identify modes via their number of multiplets. For more rapid rotation, split modes of adjacent radial orders overlap, confusing the identification [31, 38].

For core hydrogen-exhausted δ Sct stars, a dense frequency spectrum develops with mixed modes having many g-type nodes in the burned-out non-convective core. For example, in models of 4 CVn, including rotational splitting, ~350 modes are expected to be unstable for $l = 0$, 1, and 2 [31, 32].

At higher stellar masses, the SPB (high-order g mode) and β Cep (low-order p-mode) pulsations are driven by the κ effect of Fe opacity at ~200,000 K. The rapid rotation of these stars allows detection and identification of modes of l and $m \geq 3$ via line-profile variations [2].

2 Complications for Mode Identification

Some complications for mode identification include stellar activity and star spots, rotation (especially rapid rotation, and rotation that is differential in radius and latitude), combinations of modes, mode coupling, and mode visibility. There are many other complications that won't be discussed here, e.g. amplitude and frequency variations, non-sinusoidal light curves, mode trapping, B fields, eclipses, planets, unknown binary components, etc.)

Concerning the prevalence of activity and spots in light curves, Balona [5] found that 8% of Kepler A stars show periodicity of 1–2 days with a traveling feature indicating spots on a differentially rotating star. Debosscher et al. [21] show examples of several irregular and aperiodic *Kepler* light curves. There is also suspicion that much of the signal in the *Kepler* light curves below 0.5 cycles/day may be due to stellar activity.

We have learned recently [35, 40, 41] that, even for modest rotation rates, the perturbation theory approach based on spherical harmonics breaks down. Single spherical harmonic mode identifications don't apply, and new modes appear (island modes, whispering gallery modes, . .). Mode coupling also affects frequency spacings [38]. Rotation also affects pulsation mode instability. In B stars, Aprilia et al. [4] find that rotations stabilizes regtrograde modes.

Combination peaks can also complicate spectra. In KIC 9700322, 80+ frequencies were identified, but all of the frequencies were combinations of two dominant frequencies, a rotationally split quintuplet, and the rotation frequency [14, 30]. Standard models do not reproduce the period ratio of the two dominant frequencies.

Mode coupling may induce additional modes, and frequency and amplitude variations [25, 26]. Parametric resonance, where $\omega_o = \omega_1 + \omega_2$, and ω_1 and ω_2 are both stable modes, could excite the stable modes to an observable amplitude and limit the amplitude of the unstable mode ω_o. Rotational splitting increases the probability of a close resonance and may explain the absence of rotating objects among large-amplitude variables.

Mode visibility drops off in full-disk integrated photometry, and from the ground we expect it to be difficult to detect modes of $l > 2$. Space-based photometry may detect modes of higher degree than $l = 2$, even up to $l = 10$ or 12, and the visibility plateaus for $l > 3$. Balona and Dziembowski [8], Barban et al. [9] and Dasynska-Daskiewicz et al. [20] discuss visibility of high-degree modes with space-based photometry, and how to cope with this added complexity.

3 Expectations Not Met

In many aspects, our theoretical expectations for main-sequence variables have not been met, prompting new questions. Many modes expected are not observed; some modes not expected are observed (e.g., for δ Scuti/γ Dor and β Cep/SPB hybrids). The dense frequency spectrum expected for stars with H-exhausted cores is not observed (e.g., 4 CVn). Not all rotational frequency multiplets are exhibited. There is a prevalance of mode coupling and linear combinations of modes (e.g., KIC 9700322).

FG Vir is perhaps one of the most tractable and best-studied δ Sct stars [2, 15, 19, 31]. Theoretical models predicted $8 + 33 + 50 = 91 l = 0$, 1, and 2 frequencies in the observed range. Sixty-five independent frequencies were observed from ground-based photometric campaigns, showing a frequency spectrum that does not match observations. Probably there are modes with $l \geq 3$ present in the observations.

Many γ Dor/δ Sct hybrids were discovered by *Kepler* and CoRoT (see, e.g., [7, 29,44]), but theoretical models predicted hybrids to exist only in a small overlapping region of the respective instability strips. Instead, nearly all of these γ Dor and δ Sct stars appear to be hybrids. Additional driving/excitation mechanisms may be required to explain this hybrid behavior.

For 4 CVn, an evolved δ Scti star with H-exhausted core, mode identifications are available for the highest amplitude of the 18 observed non-combination modes [12, 13]) However, Guzik et al. [31, 32] find 336 $l = 0$, 1, and 2 modes are predicted in the observed frequency range, and efforts to explain the reduced number of visible modes have so far not been satisfactory.

SPB and β Cep hybrids show fewer modes than expected, and hybrids that are not predicted. Concerning the hybrid stars ν Eri and 12 Lac, some g modes are predicted in ν Eri, but not at the frequency of the observed mode, and for 12 Lac, all of the g modes are damped, so hybrid behavior is not predicted [24].

Are the missing modes of too low amplitude to be observed? Are many modes with $l > 3$ visible in space-based photometry? Do modes appear and disappear because of resonant mode coupling or stochastic excitation? Is energy insufficient to drive all of the modes? Is dissipation/damping higher than expected? What is the role of mode trapping? Why don't we see all multiplets of rotationally split modes? How do we explain the unexpected β Cep/SPB and γ Dor/δ Sct hybrids? Do we need additional driving or excitation mechanisms? Why do we see combinations of modes and mode coupling for some stars but not for others?

4 What Do We Do in Practice?

What do we do in practice? As a first step, we use observational constraints to calculate models and frequencies to compare to observations, and try to match up patterns. We search for combination frequencies of modes in the spectrum to separate them from intrinsic modes. We can use expected equal frequency or period spacing (for high radial-order p or g modes), and small and large separations, the frequency ratio method [37], histograms of period spacings, and Echelle diagrams as clues to mode identification. Scaled Echelle diagrams can be used for comparative asteroseismology, and in trying to sort out $l = 0$, 2 and $l = 1$, 3 ridges [10]. Frequency difference histograms for observed and expected frequencies may reveal a characteristic spacing even for modes without asymptotic spacing, and when including modes of higher degree [9, 27]. We can use additional constraints on luminosity and T_{eff} to constrain radial order for sets of homologous models. We can look for breaks in patterns indicating avoided crossings to help in selecting models of the right age, mass, and core structure. We can look for frequency patterns arising from rotational splitting [38]. For brighter stars and higher amplitude modes, we can use photometric and spectroscopic mode ID methods [1]. We can acquire longer time series to increase precision of frequency determinations and to measure mode lifetimes and frequency/amplitude variations.

Energy is partitioned and exchanged at various rates between many different processes in stars: gravitational, nuclear, thermal, radiation, rotational, convective, pulsation mode kinetic energy, viscous dissipation, B fields, winds, waves, etc. Understanding this balance may be important to understanding the behavior of individual stars.

Along these lines, Moya and Rodríguez-Lópes [36] estimated how much energy is needed to excite the thousands of modes observed for some CoRoT stars. They calculated mode kinetic energies and linear amplitudes for δ Sct and γ Dor star models, and scaled amplitudes to the observations of two CoRoT stars. They found that the energy per unit time needed by the modes to be overstable is 4–5 orders of magnitude less than the luminosity, and the total kinetic energy of all of the modes is 6–7 orders of magnitude less than the stellar gravitational potential energy. However, they only included modes of $l = 0$–7, and in reality there may be thousands more undetected modes present. See also Dupret et al. [22] for an additional study of energy considerations in pulsators.

5 Future Directions

To make further progress, we need to understand mode amplitudes (and phases) more fundamentally, and not rely only on frequency patterns for mode identification. Understanding amplitudes will require nonlinear, nonadiabatic, nonradial codes with more sophisticated outer boundary and model atmosphere treatments. These models may need to be two- or three-dimensional to include the effects of convection and rapid and differential rotation, and may require considerations of B fields to capture the effects of stellar activity cycles and offset magnetic and rotational axes.

We should consider stochastic excitation as a mechanism to fill in the frequency spectrum. Stochastic excitations have been considered and searched for in δ Sct [3, 17] andγ Dor [39] stars, γ Dor/δ Sct hybrids [16], and B stars [6, 11], and across the HR diagram [34, 42].

We need to focus on dissipation mechanisms, viscosity, and driving and damping balance in nonadiabatic simulations. Horizontal shear of gravity modes in interiors of evolved stars may contribute to damping and elimination of some modes [31, 32].

Nonlinear pulsation modeling can be used to study amplitude saturation and validate viscosity treatments, as was recently shown by Smolec and Moskalik [43] for β Cep models. However, their study was limited to radial modes. Mode trapping and mode kinetic energy could influence mode amplitudes and visibility. Guzik et al. [33] found that mode kinetic energy was at a minimum for the 1 c/d frequencies most commonly observed in γ Dor stars, and mode trapping was explored by Breger and Pamyatnykh [12] and Dziembowski [23] to explain the selection of the observed $l = 2$ modes of 4 CVn.

The viewgraphs for this talk and an expanded bibliography for this paper including referenced article titles are available at:
/www.mindspring.com/~jguzik/GuzikGranada2011Bibl.pdf
/www.mindspring.com/~jguzik/GuzikGranada2011Talk.pdf

References

1. Aerts, C., Christensen-Dalsgaard, J., Kurtz, D.W.: Asteroseismology, Springer Astronomy and Astrophysics Library (2010)
2. Aerts, C., Briquet, M., Carrier, F., Desmet, M., Zima, W.: Comm. Asteroseism. **157**, 29 (2008)
3. Antoci, V., et al.: Nature **477**, 570 (2011)
4. Aprilia, Lee, U., Saio, H.: MNRAS **412**, 2265 (2011)
5. Balona, L.A.: MNRAS **415**, 1691 (2011)
6. Balona, L.A., et al.: MNRAS **413**, 2403 (2011)
7. Balona, L.A., et al.: MNRAS **415**, 3531 (2011)
8. Balona, L., Dziembowski, W.A.: MNRAS **309**, 221 (1999)
9. Barban, C., et al.: ESA-SP **464**, 399 (2001)
10. Bedding, T.R., Kjeldsen, H.: Comm. Asteroseism. **161**, 3 (2010)
11. Belkacem, K., Dupret, M.A., Noels, A.: A&A **510**, A6 (2010)
12. Breger, M., Pamyatnykh, A.A.: ASP Conf. Ser. **259**, 388 (2002)
13. Breger, M., et al.: A&A **349**, 225 (1999)
14. Breger, M., et al.: MNRAS **414**, 1721 (2011)
15. Breger, M., et al: A&A **435**, 955 (2005)
16. Campante, T.I., et al.: arXiv:1003.4427v1 (2010)
17. Casas, R., et al: ApJ **697**, 522 (2009)
18. Christensen-Dalsgaard, J.: Lecture Notes on Stellar Oscillations, http://users-phys.ak.dk/jcd/oscilnotes (2003)
19. Daszyǹska-Daszkiewicz et al.: A&A **438**, 653 (2005)
20. Daszyǹska-Daszkiewicz, J., Dziembowski, W.A., Pamyatnykh, A.: Mem. S.A.It. **77**, 113 (2006)
21. Debosscher, J., Blomme, J., Aerts, C., De Ridder, J.: A&A **529**, 89 (2011)
22. Dupret, M.-A., Grigahcène, A., Thèado, S., Bourge, P.-O., Gabriel, M.: ESA SP **624** (2006)
23. Dziembowski, W.A.: In: Provost, J., Schmider, F.-X. (eds.) Sounding Solar and Stellar Interiors, p. 317. IAU, Netherlands (1997)
24. Dziembowski, W.A., Pamyatnykh, A.A.: MNRAS **385**, 2061 (2008)
25. Dziembowski, W., Królikowska, M.: Acta Astron. **35**, 5 (1985)
26. Dziembowski, W., Królikowska, M, Kosovitchev, A.: Acta Astron. **38**, 61 (1988)
27. Garciá Hernández, A., et al.: A&A **506**, 79 (2009)
28. Goupil, M.J., Dziembowski, W.A., Pamyatnykh, A.A., Talon, S.: ASPC **210**, 267 (2000)
29. Grigahcène, A., et al.: ApJL **713L**, 192 (2010)
30. Guzik, J.A., Breger, M.: Comm. Asteroseism. **162**, 62 (2011)
31. Guzik, J.A., Bradley, P.A., Templeton, M.R.: ASPC **210**, 247 (2002)
32. Guzik, J.A., Austin, B.A., Bradley, P.A., Cox, A.N.: ASPC **310**, 462 (2004)
33. Guzik, J.A., et al.: ApJL **542**, L57 (2000)
34. Houdek, G., Balmforth, N.J, Christensen-Dalsgaard, J., Gough, D.O.: A&A **351**, 582 (1999)
35. Lignières, F., Rieutord, F., Reese, D.: A&A **455**, 607 (2006)
36. Moya, A., Rodríguez-López, C.: ApJL **710**, L7 (2010)
37. Moya, A., et al.: Astron. Nachr. **329**, 541 (2008)
38. Pamyatnykh, A.A.: Astrophys. Spa. Sci. **284**, 97 (2003)
39. Pereira, T.M.D., et al.: A&A **464**, 659 (2007)

40. Reese, D., Lignières, F., Rieutord, M.: A&A **455**, 621 (2006)
41. Reese, D.R.: Astron. Nachr. **331**, 1038 (2010)
42. Samadi, R., et al.: A&A **463**, 297 (2007)
43. Smolec, R., Moskalik, P.: MNRAS **377**, 645 (2007)
44. Uytterhoeven, K., et al.: A&A **534**, 125 (2011)

Towards Precise Asteroseismology of Solar-Like Stars

A. Grigahcène, M.-A. Dupret, S.G. Sousa, M.J.P.F.G. Monteiro, R. Garrido, R. Scuflaire, and M. Gabriel

Abstract Adiabatic modeling of solar-like oscillations cannot exceed a certain level of precision for fitting individual frequencies. This is known as the problem of near-surface effects on the mode physics. We present a theoretical study which addresses the problem of frequency precision in non-adiabatic models using a time-dependent convection treatment. We find that the number of acceptable model solutions is significantly reduced and more precise constraints can be imposed on the models. Results obtained for a specific star (β Hydri) lead to very good agreement with both global and local seismic observables. This indicates that the accuracy of model fitting to seismic data is greatly improved when a more complete description of the interaction between convection and pulsation is taken into account.

1 Introduction

Two conditions are necessary for precise probing of the physics of stellar interiors with asteroseismology. One condition concerns the quality of the observed data. Space missions have resulted in data of unprecedented quality and improved data

---

A. Grigahcène (✉) · S.G. Sousa
Centro de Astrofísica da Universidade do Porto, Rua das Estrelas, 4150-762 Porto, Portugal

M.-A. Dupret · R. Scuflaire · M. Gabriel
Institut d'Astrophysique et de Géophysique, Université de Liège, Allée du 6 Août 17-B 4000 Liège, Belgium

M.J.P.F.G. Monteiro
Centro de Astrofísica da Universidade do Porto, Rua das Estrelas, 4150–762 Porto, Portugal

Departamento de Física e Astronomia, Faculdade de Ciências da Universidade do Porto, Porto, Portugal

R. Garrido
Instituto de Astrofísica de Andalucía, CSIC, PO Box 3004, 18080 Granada, Spain

Grigahcène, A. et al.: *Towards Precise Asteroseismology of Solar-Like Stars.*
Astrophys Space Sci Proc. **31**, 209–212 (2013)
DOI 10.1007/978-3-642-29630-7_39,

analysis. The other condition is that models should be able to reproduce the complex physics at work in these stars, allowing for precise interpretation of the observations.

The treatment of convection is one of the largest sources of uncertainty in stellar astrophysics. For cool pulsating stars, the convective envelope introduces severe complications for modelling the oscillations. In these stars, the interaction between pulsation and convection is very important for proper modeling of the oscillations. For example, this interaction has been found to be a major source of damping in δ Scuti stars [5] and in driving of oscillations in γ Doradus stars [4].

For stars with solar-like pulsations, it turns out that the transition region where the thermal relaxation time is of the same order as the pulsation period is located inside the superficial convective zone. Since these near-surface layers have not been properly modeled, they give rise to discrepancies between the computed and observed frequencies [6].

Gabriel's formalism of time-dependent convection (TDC) [7, 8], as implemented in the MAD code [2], leads to very important results for models of δ Sct and γ Dor stars [4, 5] and also for the solar damping rates [3].

In this work, we present TDC results for models of β Hydri, a G2 IV subgiant showing solar-like oscillations. A recent list of observed pulsation frequencies in this star is given in Brandão et al. [1]. We compare these frequencies with calculations from adiabatic models with and without the near-surface corrections of Kjeldsen et al. [9].

2 TDC Modeling of β Hydri

We constructed a grid of TDC models with equilibrium models and non-adiabatic non-radial TDC oscillations calculated as described in Dupret et al. [5]. We then searched for the best-fitting model by maximizing the likelihood:

$$\mathscr{L} = \left(\prod_{j}^{N_{\mathscr{O}}} \frac{1}{\sqrt{2\pi}\sigma_j} \right) \times \exp(-\chi^2/2),$$

where $\chi^2 = \chi_\nu^2 + \chi_{\mathscr{P}}^2 = \frac{1}{N_{\mathscr{O}}} \sum_{j=1}^{N_{\mathscr{O}}} \left(\frac{\mathscr{O}_j^{theo} - \mathscr{O}_j^{obs}}{\sigma_j^{\mathscr{O}}} \right)^2$, describes the match between the $N_{\mathscr{O}}$ parameters of the theoretical model $\mathscr{O}^{theo}$ and the observed values $\mathscr{O}^{obs}$. These parameters are the effective temperature, T_{eff}, the mass, M, the gravity, log g, the radius, R, and the chemical composition, (X, Z) as well as the frequencies of the radial modes. The quantities χ_ν^2 and $\chi_{\mathscr{P}}^2$ give the partial values of χ^2 associated with the frequencies and the fundamental parameters respectively.

The difference between the theoretical and observed frequencies is shown in Fig. 1 and the resulting values of χ_ν^2 listed in column 9 of Table 1. We note that the adiabatic models give the poorest agreement with observations. The TDC frequencies agree best with observations as shown by χ_ν^2. The fact that χ_ν^2 for models

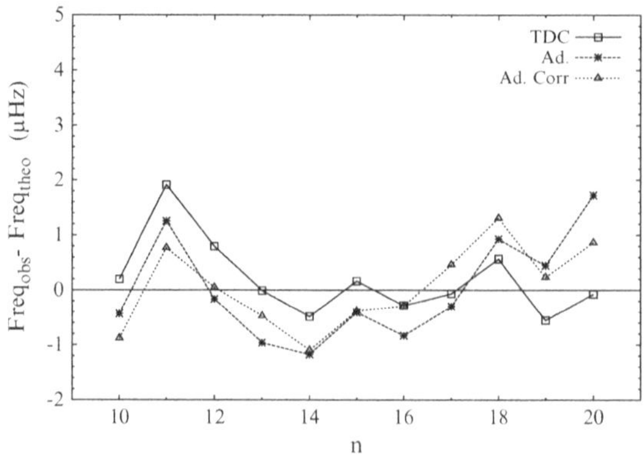

Fig. 1 The difference between theoretical and observed frequencies for radial modes using TDC (*squares*), pure adiabatic (*asterisks*) and adiabatic with near-surface corrections (*triangles*)

Table 1 Some properties of the best-fitting models

Model	M ($M_\odot$)	R($R_\odot$)	Age (Gyr)	T_{eff} (K)	Z/X	α_{MLT}	$\langle \Delta\nu \rangle$ (μ Hz)	χ^2_ν	$\chi^2_{\mathscr{P}}$
TDC	1.072	1.8211	8.1957	5858	0.0194	2.300	57.43	0.14	0.26
Adiab.	1.070	1.8223	8.2923	5856	0.0194	2.500	57.67	0.58	0.29
Corr.	1.080	1.8207	7.8221	5856	0.0194	1.900	57.63	0.40	0.31

with near-surface corrections is much larger than for TDC models shows that near-surface corrections only apply to stars very similar to the Sun. This is, of course, to be expected since the Sun is used to obtain these corrections.

In Fig. 2 we give the probability distribution as a function of mass and age. In Table 1, columns 2–7 give the derived global parameters. Column 10 lists values of $\chi^2_{\mathscr{O}}$. The adiabatic models give the largest radii and the lowest masses. TDC models and models with near-surface corrections give different masses. Models with near-surface corrections give the lowest ages because they use the smallest values of the mixing-length parameter, α_{MLT}. The pure adiabatic models give the highest ages for the same reason: they use the highest values of α_{MLT}. Column 8 in Table 1 gives the average large separations, $\langle \Delta\nu \rangle$. Note that these are all very similar and differ by less than 1 μHz from the observed value of 57.48 μHz. The large separation obtained using TDC agrees best with the observed value.

On the other hand the mixing length, α_{MLT}, which is a free parameter, differs considerably for the different models and is higher than the calibrated value for the Sun (α_{MLT} ~1.8). This agrees with the expected variability of α_{MLT} across the HR diagram.

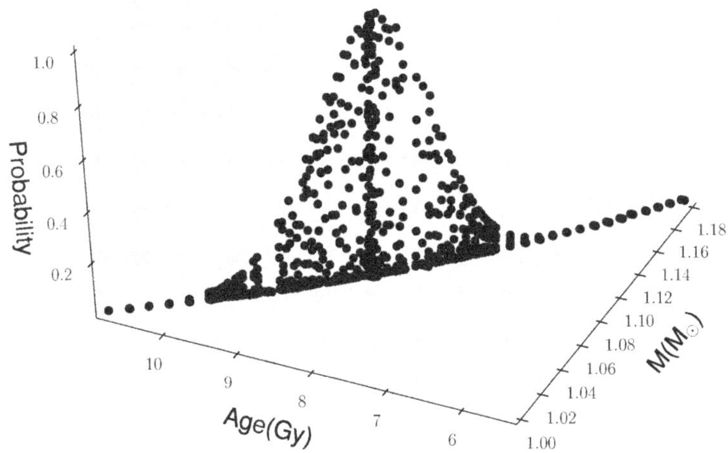

Fig. 2 The relative probability distribution as a function of mass, M, and age

3 Conclusions

The main results of this work are as follows. (a) TDC leads to a significant improvement in the agreement between calculated and observed frequencies of solar-like oscillations as compared to the pure adiabatic frequencies. (b) A physically more robust modeling of the mode physics near the surface provides better constraints on the global and free stellar parameters. (c) Near-surface corrections are probably only valid for stars most closely resembling the Sun.

References

1. Brandão, I.M., Doğan, G., Christensen-Dalsgaard, J., et al.: A&A **527**, 37 (2011)
2. Dupret, M.A.: A&A **366**, 166 (2001)
3. Dupret, M.A., Barban, C., Goupil, M.-J., Samadi, R., Grigahcène, A., Gabriel, M.: In: Proceedings of SOHO 18/GONG 2006/HELAS I. Beyond the spherical Sun, vol. 624, p. 97 (2006)
4. Dupret, M.-A., Grigahcène, A., Garrido, R., et al.: A&A **414**, L17 (2004)
5. Dupret, M.-A., Grigahcène, A., Garrido, R., et al.: A&A **435**, 927 (2005)
6. Dziembowski, W.A., Paterno, L., Ventura, R.: A&A **200**, 213 (1988)
7. Gabriel, M.: Bull. Astron. Soc. India **24**, 233 (1996)
8. Grigahcène, A., Dupret, M.-A., Gabriel, M., et al.: A&A **434**, 1055 (2005)
9. Kjeldsen, H., Bedding, T.R., Christensen-Dalsgaard, J.: ApJL **683**, L175 (2008)

Part V
Oscillations Versus Activity Plus Planet Transits

Current State of the Modelling of Photospheric Activity

A.F. Lanza

Abstract I briefly review the current state of the modelling of photospheric activity based on the high-precision optical light curves obtained with MOST, CoRoT, and Kepler. These models can be used to search for active longitudes where activity is preferentially concentrated, estimate the amplitude of stellar differential rotation, and look for short-term activity cycles as, e.g., in the case of CoRoT-2. In the case of a late-type star accompanied by a transiting hot Jupiter, the small light modulations observed during transits when a dark spot is occulted by the disc of the planet are also briefly considered. They can be used to derive information on individual active regions as well as on stellar rotation and the spin–orbit alignment of the system.

1 Introduction

In the Sun, we can study the photosphere in detail and map its brightness inhomogeneities associated with convective motions, oscillations, and magnetic fields. In particular, magnetic fields give rise to solar *active regions* characterized by dark spots, bright faculae and an enhanced network.

The transit of an active region across the solar disc as our star rotates produces a characteristic modulation of the optical flux with a rise when the region appears close to the approaching limb, a dip when it is closest to the central meridian, and another rise when it reaches the receding limb. The flux rises are associated with faculae that are brighter close to the limb and do not show an appreciable contrast when

A.F. Lanza (✉)
INAF-Osservatorio Astrofisico di Catania, via S. Sofia, 78, 95125 Catania, Italy
e-mail: nuccio.lanza@oact.inaf.it

Lanza, A.F.: *Current State of the Modelling of Photospheric Activity.*
Astrophys Space Sci Proc. **31**, 215–220 (2013)
DOI 10.1007/978-3-642-29630-7_40,

they are close to the disc centre, while the flux dip is produced by the dark sunspots whose photometric effect is maximum when their projected area is maximum, i.e., when the active region is closest to the centre of the disc (cf., e.g., [6]).

The collective behaviour of solar active regions shows a remarkable regularity. The total sunspot area oscillates with a mean period of $\sim$ 11 year that is the main feature of the solar *activity cycle*. The latitudes where sunspot groups form migrate along the cycle, with the first groups appearing at $\sim$ 30°–35° and then migrating towards the equator as the cycle progresses. At the maximum of activity, sunspots form at a latitude of $\sim$ 15°–20° to reach a minimum latitude of 5°–10° at the end of the cycle, when the sunspot groups associated with the next cycle begin to appear at high latitudes.

The variation of the sunspot area is accompanied by a variation of the eigenfrequencies of the solar p-mode oscillations and of the amplitude and full width at half maximum (hereafter FWHM) of their peaks in the power spectrum. Since we wish to compare the solar behaviour with that of distant stars, we focus on low-degree modes ($\ell = 0, 1, 2$) that can be detected in disc-integrated observations. The variation of their eigenfrequencies along the previous solar cycle 23 and the onset of the present solar cycle 24 has been reported by [26], while [10] have presented the variation of the amplitude and FWHM of the mode peaks in the power spectrum. Note that the onset of the new cycle 24 has been apparent in the $\ell = 0$ and 2 modes, while it has not been detected in the $\ell = 1$ modes. This is related to the spatial distribution of the magnetic perturbations in the Sun because the $\ell = 0, 2$ modes are more sensitive to the high-latitude perturbations, while the $\ell = 1$ modes respond mainly to the low-latitude perturbations. Since the magnetic fields are localized at high latitudes at the beginning of a new cycle, their effect can be detected in the $\ell = 0, 2$ modes even before the first sunspot groups appear (cf. [26] for details).

Analogous phenomena are expected in late-type stars with a level of magnetic activity comparable with the Sun. They have outer convection zones that host hydromagnetic dynamos generally leading to cycles [2] and display convection-driven p-mode oscillations. Indeed, [7] have detected a dependence of the eigenfrequencies and the mode amplitudes on the level of magnetic activity in the F-type star HD 49933, recently observed by the CoRoT (Convection, Rotation and Transits) space telescope.

Taking the Sun as a template, to perform a comparative study of the impact of activity on p-mode oscillations in distant stars, we need to derive: (a) the variation of the total spotted area vs. time, to trace stellar activity cycles; (b) the latitude of the active belts vs. the phase of the activity cycle, assuming that there is a regular migration as in the Sun (see [19] for a detailed discussion of the possible behaviours). In the next sections, we shall introduce the methods to map stellar photospheres that allow us to achieve these goals and present some application to the case of CoRoT targets. The advantages of stars accompanied by transiting hot Jupiters will be also emphasized.

2 Methods to Map the Photospheres of Late-Type Stars

To map stellar photospheres, we can apply methods based on spectroscopic and/or photometric data. Doppler Imaging (hereafter DI) is the most powerful technique based on spectroscopy, while photometric spot modelling is based on fitting wide-band light curves.

DI is based on the line profile distortions induced by surface brightness inhomogeneities that can be resolved when the stellar rotation is the main source of line broadening [32]. In practice, this implies $v \sin i \geq 10$–15 km s^{-1} in a main-sequence star. In G or K-type dwarf stars, this corresponds to a maximum rotation period of $\sim$ 4–5 days that implies an optical variability induced by magnetic activity of at least 0.02–0.05 mag. This may hamper the detection of p-mode oscillations as recently found by [3] by analysing a sample of Kepler targets (see also [20] for the case of the active CoRoT target HD 175726). Another limitation of DI is the lack of sensitivity to a spot distribution that is uniform on length scales comparable with or larger than its spatial resolution. This poses a severe limitation on the measurement of the total spotted area in stars that are slowly rotating, as most of the asteroseismic late-type targets. Nevertheless, DI is capable of retrieving the latitude of starspots which is not possible with spot modelling, except in very favourable circumstances, and can be applied to spectropolarimetric observations of magnetic sensitive lines to map the surface magnetic fields and look for magnetic cycles (Zeeman Doppler Imaging, see, e.g., [5]).

Spot modelling techniques are based on the reproduction of the rotational modulation of the optical flux induced by starspots. If there is a single spot on a rotating star, the time of light minimum can be used to derive its longitude and the maximum deficit of the flux gives a measure of the projected area of the spot. If the star is not viewed equator-on, the duration of the spot transit across the stellar disc can be used to estimate its latitude. As a matter of fact, several spots are usually present on the surface of an active star which makes the light curve inversion non-unique and often unstable because small changes in the data imply large changes in the derived spot positions and unprojected areas. In principle, a light curve is a one-dimensional data set, thus the derivation of a two-dimensional map of the photosphere without some additional assumptions is impossible. Nevertheless, the longitude of the spots and their total projected area are robust quantities that can usually be retrieved unambiguously from a high-precision light curve. This has been demonstrated in the case of the Sun, for which we have spatially resolved maps of the photosphere, by e.g., [12].

To alleviate the non-uniqueness and instability problems, two approaches have been applied. The first adopts a limited number of spots, usually up to three or four, and fits a limited number of free parameters including the spot coordinates and areas *(few-spot modelling)*. Several works have discussed the technique beginning from the earliest in the 1970 and 1980 [24, 31], to the first application to time series of data covering several consecutive rotations [30], and finally to the time-series photometry obtained by space-borne telescopes like MOST (the Microvariability and Oscillation of Star satellite, [25]).

The other approach to spot modelling consists in assuming a continuous distribution of small spots, parameterized by a filling factor f giving the fraction of the surface occupied by the spots at each given location. To avoid the instability and non-uniqueness of the solution, the method needs a priori assumptions on the distribution of f to regularize the solution. The best approach, in the sense that it reproduces the distribution of sunspot groups in the Sun [12], is based on the *Maximum Entropy regularization* (hereafter ME), where we look for the map with the maximum Shannon entropy among the infinite distributions of the filling factor that fit the light curve down to the precision of the measurements. From ME maps, the distribution of the filling factor vs. longitude and the variation of the total spotted area vs. time are derived. The information on spot latitude is usually very limited or absent in a light curve (cf. e.g., [13]), so we should not attempt to retrieve it from ME maps.

Finally, we should mention the possibility of using the occultations of the starspots by a transiting planet as a high-pass spatial filter to map the fine structure of the spot distribution within the occulted band. The method has been successfully applied to, e.g., CoRoT-2 by [28, 29] and provides information on the characteristics of individual spots as well as on stellar differential rotation and the spin–orbit alignment of the planetary system [22].

3 Results

Long-term photometry allows us to derive stellar butterfly diagrams by tracing the systematic variation of the stellar rotation period as derived from the optical flux modulation vs. the time. The variation of the total spotted area can be measured simultaneously by, e.g., the variation of the mean light level from one season to the next, allowing us to trace the phase of the cycle and, in combination with the variation of the rotation period, the direction of migration of the spots [18]. It is interesting to note that in some stars the activity belts may migrate polewards (see, e.g., [19]). In stars that rotate as slowly as the Sun, it is difficult to measure the rotation period using photospheric spots because they evolve on time scales of 10–15 days in the Sun, i.e., significantly shorter than the rotation period itself. Chromospheric faculae live longer and are better tracers to look for the systematic variations associated with the solar cycle (see [9] and references therein).

An estimate of the starspot temperatures based on multi-wavelength ground-based photometry has been presented by [1]. The high-precision space-borne photometry provided by MOST has allowed to measure the surface differential rotation of several late-type stars [4]. Among the latest results, those obtained by [21] are particularly interesting from the point of view of asteroseismology because they refer to F-type main-sequence stars observed by CoRoT to look for p-mode oscillations. In addition to the spot lifetimes, they estimate the amplitude of the surface differential rotation and the inclination of the spin axis of the star to the line of sight with a typical uncertainty of 20°–30°. Asteroseismology can do much better providing us with an inclination with an uncertainty of 5°–10° in the case of such stars that rotate 4–5 times faster than the Sun [8]. In turn, this can be used to improve the estimate of the differential rotation from the spot modelling.

The activity of several planet-hosting stars discovered by CoRoT has been investigated. CoRoT-2 shows a short-term activity cycle with a period of about 29 days that could be analogous to the solar Rieger cycles [17, 23, 33] or may be induced by some kind of star-planet interaction [11, 13]. Active longitudes have been detected and estimates of the surface differential rotation have been obtained for the other targets CoRoT-4 [14], CoRoT-7 [15], and CoRoT-6 [16].

The power of starspot occultation during transits in active stars accompanied by hot Jupiters has been shown by [27] in the case of HAT-P-11 as observed by the Kepler space telescope. In addition to a quite accurate estimate of the inclination of the stellar spin, they have obtained a measurement of the true misalignment of the system by combining spot modelling with spectroscopic measurements of the Rossiter–McLaughlin effect. The latitude of the starspots has been constrained thanks to the large misalignment of the system. This offers a new opportunity to build stellar butterfly diagrams by means of the long-term Kepler photometry.

4 Discussion and Conclusions

The recent detection of a dependence of the p-mode properties on the activity level in HD 49933, similar to that observed in the Sun, has revived the interest in studying stellar activity in late-type stars with outer convection zones. Those stars display p-mode oscillations excited by convection as well as activity phenomena produced by dynamo-generated magnetic fields. We have illustrated as the same high-precision photometric data collected to measure p-mode oscillations (and in several cases to search for planetary transits) can be used to model the distribution of starspots vs. longitude in a stellar photosphere and trace the variation of their area and migration along the activity cycle. This will allow us a better understanding of the impact of magnetic activity on p-mode oscillations in solar-like stars thanks to the long-term photometry provided by the CoRoT and Kepler missions (see the talk by S. Mathur). Spot modelling can significantly benefit from the determination of the stellar inclination made possible by asteroseismology. This in turn improves the estimation of the stellar differential rotation which is of fundamental importance to understand the rotational splitting of the p-mode frequencies and the operation of the stellar dynamo. Stars accompanied by a transiting planet offer new opportunities for the spot modelling thanks to the occultation of the spots by the planet and are also extremely interesting asteroseismic targets (see the talk by A. Moya).

Acknowledgements The author wishes to thank the SOC of the 20th Stellar Pulsation Conference for their kind invitation to attend the Conference and their nice hospitality in Granada, and Prof. B. Mosser for interesting discussions.

References

1. Amado, P.J., Zboril, M.: A&A **381**, 517 (2002)
2. Baliunas, S.L., Donahue, R.A., Soon, W.H., et al.: ApJ **438**, 269 (1995)
3. Chaplin, W.J., Bedding, T.R., Bonanno, A., et al.: ApJL **732**, L5 (2011)
4. Croll, B., Walker, G.A.H., Kuschnig, R., et al.: ApJ **648**, 607 (2006)
5. Fares, R., Donati, J.-F., Moutou, C., et al.: MNRAS **398**, 1383 (2009)
6. Fröhlich, C., Lean, J.: Astron. Astrophys. Rev. **12**, 273 (2004)
7. García, R.A., Mathur, S., Salabert, D., et al.: Science **329**, 1032 (2010)
8. Gizon, L., Solanki, S.K.: ApJ **589**, 1009 (2003)
9. Hempelmann, A., Donahue, R.A.: A&A **322**, 835 (1997)
10. Jiménez-Reyes, S.J., García, R.A., Jiménez, A., Chaplin, W.J.: ApJ **595**, 446 (2003)
11. Lanza, A.F.: Ap&SS **336**, 303 (2012)
12. Lanza, A.F., Bonomo, A.S., Rodonò, M.: A&A **464**, 741 (2007)
13. Lanza, A.F., Pagano, I., Leto, G., et al.: A&A **493**, 193 (2009a)
14. Lanza, A.F., Aigrain, S., Messina, S., et al.: A&A **506**, 255 (2009b)
15. Lanza, A.F., Bonomo, A.S., Moutou, C., et al.: A&A **520**, A53 (2010)
16. Lanza, A.F., Bonomo, A.S., Pagano, I., et al.: A&A **525**, A14 (2011)
17. Lou, Y.-Q.: ApJ **540**, 1102 (2000)
18. Messina, S., Guinan, E.F.: A&A **409**, 1017 (2003)
19. Moss, D., Sokoloff, D., Lanza, A.F.: A&A **531**, A43 (2011)
20. Mosser, B., Michel, E., Appourchaux, T., et al.: A&A **506**, 33 (2009a)
21. Mosser, B., Baudin, F., Lanza, A.F., et al.: A&A **506**, 245 (2009b)
22. Nutzman, P.A., Fabrycky, D.C., Fortney, J.J.: ApJL **740**, L10 (2011)
23. Oliver, R., Ballester, J.L., Baudin, F.: Nature **394**, 552 (1998)
24. Rodonó, M., Cutispoto, G., Pazzani, V., et al.: A&A **165**, 135 (1986)
25. Rucinski, S.M., Walker, G.A.H., Matthews, J.M., et al.: PASP **116**, 1093 (2004)
26. Salabert, D., García, R.A., Pallé, P.L., Jiménez-Reyes, S.J.: A&A **504**, L1 (2009)
27. Sanchis-Ojeda, R., Winn, J.N. (2011) arXiv:1107.2920
28. Silva-Valio, A., Lanza, A.F.: A&A **529**, A36 (2011)
29. Silva-Valio, A., Lanza, A.F., Alonso, R., Barge, P.: A&A **510**, A25 (2010)
30. Strassmeier, K.G., Bopp, B.W.: A&A **259**, 183 (1992)
31. Torres, C.A.O., Ferraz Mello, S.: A&A **27**, 231 (1973)
32. Vogt, S.S., Penrod, G.D.: PASP **95**, 565 (1983)
33. Zaqarashvili, T.V., Carbonell, M., Oliver, R., Ballester, J.L.: ApJ **709**, 749 (2010)

Pulsating Stars Harbouring Planets

A. Moya

Abstract Why bother with asteroseismology while studying exoplanets? There are several answers to this question. Asteroseismology and exoplanetary sciences have much in common and the synergy between the two opens up new aspects in both fields. These fields and stellar activity, when taken together, allow maximum extraction of information from exoplanet space missions. Asteroseismology of the host star has already proved its value in a number of exoplanet systems by its unprecedented precision in determining stellar parameters. In addition, asteroseismology allows the possibility of discovering new exoplanets through time delay studies. The study of the interaction between exoplanets and their host stars opens new windows on various physical processes. In this review I will summarize past and current research in exoplanet asteroseismology and explore some guidelines for the future.

1 Introduction

In a review devoted to the field of study of pulsating stars harbouring planets, Vauclair [42] wondered "why bother with asteroseismology in exoplanet research?". She gave several answers to this question. This review is based on further elaboration of these topics. In addition, I will discuss recent developments which relate to this question.

The potential of asteroseismology for the study of exoplanetary systems was already evident soon after the discovery of the first exoplanets. However, it is

---

A. Moya
Departamento de Astrofísica, Centro de Astrobiología (INTA-CSIC), PO BOX 78, 28691 Villanueva de la Cañada, Madrid, Spain
e-mail: amoya@cab.inta-csic.es

Moya, A.: *Pulsating Stars Harbouring Planets.*
Astrophys Space Sci Proc. **31**, 221–226 (2013)
DOI 10.1007/978-3-642-29630-7_41,

only in the last six years that the interaction between exoplanet research and asteroseismology began to show promising results. Let us begin by investigating some of the topics that resulted from Vauclair's question.

2 Synergy: (Almost) the Same Instruments are Used

Asteroseismology and exoplanet detection are both based on the analysis of time series. Currently, the best way of detecting exoplanets is through radial velocity measurements (ground based observations) or the transit technique which demands a high photometric precision, mainly attained from space missions. Asteroseismology has the same requirements. Most of this review is devoted to the contribution of asteroseismology to exoplanet research. The contribution of exoplanet science to asteroseismology cannot be neglected since exoplanets may offer an explanation for some asteroseismological puzzles.

The synergy between the two fields will always co-exist in present and future exoplanet space missions. It is in the characterization of the host star that asteroseismology plays a key role [9, 18, 30].

3 Noise: One Person's Noise is Another Person's Signal

Historically, the exoplanet community has treated the variability of the host star, whether activity or pulsation, as noise which makes the detection of exoplanets very difficult [10]. Attempts have been made to reduce this "noise" using different techniques. However, sometimes the noise is a signal that can assist in the detection of the exoplanet and characterization of the central star. This is the idea behind the Plato Data Center, where light curves will be stored and studied by a collaboration between the exoplanet and asteroseismological communities.

4 Characterization: Obtain Precise Values of the Parameters of Exoplanets-Host Stars

The study and characterization of stars hosting planets is an essential requirement in understanding the physics of exoplanets. This can be illustrated by a summary of the most successful techniques that are presently available for the detection of exoplanets. Currently, five main techniques are in use, but by far the most common are radial velocities, transits and direct imaging.

- *Radial velocities.* This technique uses the periodic variation in the radial velocity due to the gravitational perturbation of the orbiting planet. This method provides a lower limit on the mass ratio.

- *Transits.* An orbiting planet crossing the line of sight results in a shallow dip in the stellar light curve. Analysis of this eclipse gives the orbital parameters and ratio of the radii.
- *Direct imaging.* This is the direct detection of a planet in an image which allows the luminosity of the planet to be determined. The luminosity of the planet and modelling the stellar age allows the mass of the planet to be determined.

It is clear that uncertainty in the general physical characteristics of the central star leads directly to uncertainty in the physical characteristics of the exoplanet [37]. There are a number of papers which discuss the use of asteroseismology for precise characterization of stellar parameters [1, 13, 19, 27, 29, 38, 39]. Most of these studies are for single main-sequence stars similar to the Sun.

All these studies stress the importance of asteroseismology. For solar-like stars the radius can be estimated with an uncertainty of about 3%, the mass to about 5% and the age to 5–10%. However, accuracy is not the same as precision; these uncertainties depend entirely on our understanding of stellar structure and evolution of the particular star.

Have we achieved the predicted precision in these parameters? A number of studies concerned with modeling of the observational data are available [6, 24]. These studies show that the stellar radius is estimated to better than 1% for main-sequence solar-like stars. The mass is estimated with a precision of about 5% and the mean density to $<0.2\%$. Currently, there are fifteen papers in which pulsating stars with planets are studied [2, 3, 7, 8, 14–16, 20, 21, 25, 28, 32]. Most of these studies concern solar-like stars or post-MS stars with solar-like pulsations. Not all these papers discuss the pulsational characteristics of the host star in the same detail. Some highlights of these studies are itemized below.

- *μ Arae.* This was the first star where asteroseismology was applied to exoplanet research. The system consists of a giant exoplanet orbiting a solar-like star [3, 5, 34, 35].
- *ι Horologii.* This is another giant exoplanet orbiting a solar-like star [4, 21, 43]. The stellar age and He content estimated from asteroseismic analysis indicates membership of the Hyades cluster. This result led to a revision of the stellar mass and therefore to the mass of the planet.
- *KIC 9904059.* Asteroseismology of this star shows that a suspected planet is in fact a companion star [15].
- *HAT-P-7* [7], *HD 46375* [14], *HD17156* [28]. In these studies, stellar characterization using asteroseismology attained the predicted precision and confirmed the parameters found in previous studies.
- *HR 8799.* This is the only pulsating star with a planetary system that has been directly imaged [26]. The age of the system and the mass of the central star are the key parameters in this case. HR 8799 is a main-sequence λ Bootis star, which renders standard techniques of age determination not very reliable. The pulsations in HR 8799 offer unique opportunities for further study of this interesting system.

It should be borne in mind that all these characterizations are very model dependent [22]. A better understanding of stellar physics is required to fully assess these predictions.

5 Planet Detection Using the Timing Method

Asteroseismology can be also used to discover planets using the timing method [33] which requires an accurate astronomical clock. The general requirements for stellar pulsation modes to be used in this way are (1) very stable pulsational periods and amplitudes, (2) several independent high amplitude modes, and (3) the periods must be short. Some types of pulsating stars that may fulfill these requirements are SdB pulsators, roAp and δ Scuti stars. For example, a planet has been found in the SdB star V391 Peg by this method [32]. The *EXOTIME* international consortium is devoted to the use of the timing method for detecting planets around SdB stars [23, 31].

6 Interaction Between Star and Planet

In addition, there are a number of on-going projects concerning star - planet interactions. For example, we know that most of the exoplanets so far discovered have metal-rich host stars, but why this is the case is not known [36]. We do not know whether the over-abundance of metals is intrinsic to the star or a result of accretion [3]. Stars with planets seem to have a lower Li abundance than stars without planets [17]. It is possible that Li depletion may be due to angular momentum transport and additional mixing caused by the interaction between the star and planet(s) [11]. Another possibility is an accretion process [40]. Asteroseismology may well shed light on these problems.

An asteroseismic study of HR 8799 has shown, for the first time, a misalignment between the rotation axis of the host star and the orbital plane of the debris disk. In addition, there is misalignment with the orbital plane of the three known planets [44].

Finally, as already demonstrated for KIC 9904059, asteroseismology can assist in uncovering false detections of exoplanets [41].

7 Future Prospects

The European Space Agency, in its "Cosmic Vision" plan, proposes four main objectives which are presented as questions. The first question is: what are the conditions for planet formation and the emergence of life? To answer this question, an advisory team suggested the following three steps: (1) discovering exoplanets, (2) characterization of these systems, and (3) the search for bio-markers [12]. As I have described in this review, asteroseismology is an important tool in planet detection and is essential for characterization of the system.

Future efforts will concentrate in the accurate characterization of host stars, especially for the brighter stars. Attempts should be made to discover more planetary systems around other types of pulsating stars and the use of asteroseismology to recognise false detections. The study of the interaction between star, planets and debris disk should be intensified. Finally, we need to apply asteroseismology to systems with exoplanets around very low mass stars and brown dwarfs.

To facilitate these objectives, there are a number of current and future ground-based instruments (HARPS, HARPS-N, CARMENES, ESPRESSO, etc.) which have been specifically designed for exoplanet studies and may also be used for asteroseismology. Space missions such as *MOST, CoRoT, Kepler*, and *BRITE* have already demonstrated the extraordinary synergy between exoplanet research and asteroseismology. Three additional space missions on exoplanets or asteroseismology are currently proposed (*Plato, EcHO, TESS*), but only some of these are likely to be accepted. There is no doubt that further space missions are required to answer the questions listed above.

Acknowledgements The author acknowledges the funding of AstroMadrid (CAM S2009/ESP-1496). This research has been funded by the Spanish grants ESP2007-65475-C02-02, AYA 2010-21161-C02-02.

References

1. Basu, S., Chaplin, W.J., Elsworth, Y.: ApJ **710**, 1596 (2010)
2. Ballot, J., et al.: A&A **530**, 97 (2011)
3. Bazot, M., et al.: A&A **440**, 615 (2005)
4. Boisse, I., et al.: A&A **528**, 4 (2011)
5. Bouchy, F., et al.: A&A **440**, 609 (2011)
6. Chaplin, W.J., et al.: Ap **713**, 169 (2010)
7. Christensen-Dalsgaard, J., et al.: ApJ **713**, 164 (2010)
8. Collier Cameron, A., et al.: MNRAS **407**, 507 (2010)
9. CoRoT, http://en.wikipedia.org/wiki/COROT (2012)
10. Dumusque, X., et al.: A&A **525**, 140 (2011)
11. Eggenberger, P., Maeder, A., Meynet, G.: A&A, **519**, 2 (2010)
12. Exoplanet Roadmap Advisory Team (EPR-AT), http://sci.esa.int/eprat (2012)
13. Gai, N., et al.: ApJ **730**, 63 (2011)
14. Gaulme, P., et al.: A&A **524**, 47 (2010)
15. Gilliland, R.L., et al.: PAS **122**, 131 (2010)
16. Hatzes, A.P., Zechmeister, M.: JPhCS **118**, 2016 (2008)
17. Israelian, G., et al.: Nature **462**, 189 (2009)
18. Kepler, http://kepler.nasa.gov/ (2012)
19. Kjeldsen, H., Bedding, T.R., Christensen-Dalsgaard, J.: IAU Symp. **253**, 309 (2009)
20. Koen, C., et al.: MNRAS **344**, 1250 (2003)

21. Laymand, M., Vauclair, S.: A&A **463**, 657 (2007)
22. Lebreton Y.: The 61st Fujihara seminar – Progress in solar/stellar physics with helio-and asteroseismology. In: Shibahashi, H. (ed.) ASP Proceedings (2012, in press). arXiv:1108.6153
23. Lutz, R., Schuh, S., Silvotti, R.: AIPC **1331**, 155 (2011)
24. Metcalfe, T.S., et al.: ApJ **723**, 1583 (2010)
25. Moya, A., et al.: MNRAS **406**, 566 (2010)
26. Moya, A., et al.: MNRAS **405**, 81 (2010)
27. Mulet-Marquis, C., et al.: A&A **506**, 153 (2009)
28. Nutzman, P., et al.: ApJ **726**, 3 (2011)
29. Quirion, P.O., Christensen-Dalsgaard, J., Arentoft, T.: ApJ **725**, 2176 (2010)
30. Plato, http://sci.esa.int/science-e/www/area/index.cfm?fareaid=104 (2012)
31. Schuh, S., et al.: Ap&SS **329**, 231 (2010)
32. Silvotti, R., et al.: Nature **449**, 189 (2007)
33. Silvotti, R., et al.: AIPC **1331**, 133 (2011)
34. Soriano, M., Vauclair, S.: JPhCS **118**, 2072 (2008)
35. Soriano, M., Vauclair, S.: A&A **513**, 49 (2010)
36. Sousa, S.G., et al.: A&A **533**, 141 (2011)
37. Southworth, J.: MNRAS **417**, 2166 (2011)
38. Stello, D., Kjeldsen, H., Bendding, T.R.: ASPC **366**, 247S (2007)
39. Stello, D., et al.: ApJ **700**, 1598 (2009)
40. Théado, S., Vauclaire, S.: Astrophys. J. **744**, 123 (2012)
41. Tingley, B., Bonomo, A.S., Deeg, H.J.: ApJ **726**, 112 (2011)
42. Vauclair, S.: Les Houches Winter School, arXiv:0809.0249 (2008)
43. Vauclair, S., et al.: A&A **482**, 5 (2008)
44. Wright, D., et al.: ApJ **278**, 20 (2011)

Successful Asteroseismology for a Better Characterization of the Exoplanet HAT-P-7b

M. Oshagh, A. Grigahcène, O. Benomar, M.-A. Dupret, M. J. P. F. G. Monteiro, R. Scuflaire, and N. C. Santos

Abstract It is well known that asteroseismology is the unique technique permitting the study of the internal structure of pulsating stars using their pulsational frequencies, which is per se very important. It acquires an additional value when the star turns out to be a planet host. In this case, the asteroseismic study output may be a very important input for the study of the planetary system. With this in mind, we use the large time-span of the *Kepler* public data obtained for the star system HAT-P-7, first to perform an asteroseismic study of the pulsating star using Time-Dependent-Convection (TDC) models. Secondly, we make a revision of the planet properties in the light of the asteroseismic study.

1 Introduction

Kepler observations are the perfect example of synergy between two separate fields in Astrophysics. This space-mission initially designed to characterize extra-solar planets is supplying the asteroseismology community with unprecedentedly precise

---

M. Oshagh (✉) · M.J.P.F.G. Monteiro · N.C. Santos
Centro de Astrofísica da Universidade do Porto, Rua das Estrelas, 4150-762 Porto, Portugal

Departamento de Física e Astronomia, Faculdade de Ciências da Universidade do Porto, Porto, Portugal
e-mail: moshagh@astro.up.pt

A. Grigahcène
Centro de Astrofísica da Universidade do Porto, Rua das Estrelas, 4150–762 Porto, Portugal

O. Benomar
Sydney Institute for Astronomy (SIfA), School of Physics, University of Sydney, NSW 2006, Australia

M.-A. Dupret · R. Scuflaire
Institut d'Astrophysique et de Géophysique, Université de Liège, Allée du 6 Août 17-B 4000 Liège, Belgium

Oshagh, M. et al.: *Successful Asteroseismology for a Better Characterization of the Exoplanet HAT-P-7b*. Astrophys Space Sci Proc. **31**, 227–230 (2013)
DOI 10.1007/978-3-642-29630-7_42,

data. The high photometric accuracy reached by *Kepler* as required to detect planet transits [5] is actually harnessed to undertake asteroseismic studies of great interest in stellar Physics. This implies a better characterization of pulsating stars that harbor planetary systems and consequently tighter constraints on the properties of the latter.

Among the 150,000 stars in the *Kepler* field of view a few of them have been identified as planet hosting stars (TrES-2 [7], HAT-P-7 [8], and HAT-P-11 [2]) previously to the launch of the satellite. Christensen-Dalsgaard et al. 2010 gave an initial asteroseismic analysis of *Kepler* data for these stars [4].

Here we give the preliminary results of our seismic study of HAT-P-7 using the *Kepler* public data (Quarters: Q0-Q2). After a precise and careful cleaning of the light-curves from the transits, we obtain the pulsation frequencies. These frequencies are used as input for a TDC modeling of HAT-P-7 oscillations. In turn, the obtained results are used to characterize the orbiting planet.

2 Kepler Data Processing and Frequency Analysis

In this study, we have used public short cadence data gathered by *Kepler* for HAT-P-7. The light-curve spans over 144 days, corresponding to the quarters Q0-Q2.[1] An effective asteroseismic exploitation of the light curve needs a careful cleaning of the 59 observed transits without loosing the structures inside the transits that may be caused by the star pulsations. Mandel and Agol [6] presented a model to produce synthetic transits in photometric light curves. Their method takes into account the stellar and planetary parameters (e.g. star and planet radii, semi-major axis, orbital inclination, stellar limb darkening coefficients) to produce the synthetic transits. It is usually used to constrain the planet parameters.

For our purpose of cleaning HAT-P-7 light curve, we used this method to produce synthetic transits corresponding to the ones observed in HAT-P-7 light curve. We used stellar and planetary parameters reported in [9]. We then subtracted the obtained synthetic transits from the observed ones. We ended up with the cleaned light-curve to be used in the frequency analysis. Figure 1 gives an example of observed and processed data.

The power spectrum of HAT-P-7, shown in the left panel of Fig. 2, has been fitted within a Bayesian framework using a maximum a posteriori approach [3]. Here, the main formulated assumption is that frequencies vary smoothly as a function of the radial order n. This allows us to improve the robustness and reliability of the extracted frequencies significantly, compared to a classical maximum likelihood approach [1].

Using the obtained frequencies and the atmospheric parameters of HAT-P-7 given by [8], we search the best model in a grid of theoretical models. We follow the same methodology as in Grigahcène et al. (these proceedings).

[1]The data are accessible through the Multi-mission Archive at STScI at http://archive.stsci.edu/.

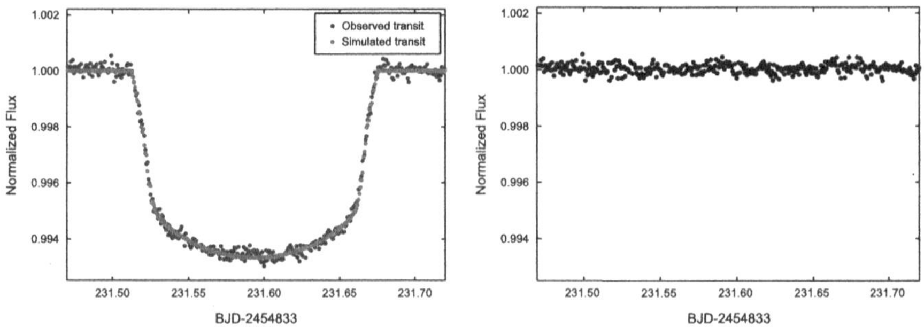

Fig. 1 *Left*: a zoom-in of one of the transits observed in HAT-P-7 light-curve obtained by Kepler. Original observed (*blue*) and synthesized transit (*red*) points are shown. See text for more details. *Right*: The same light-curve after removing the transit

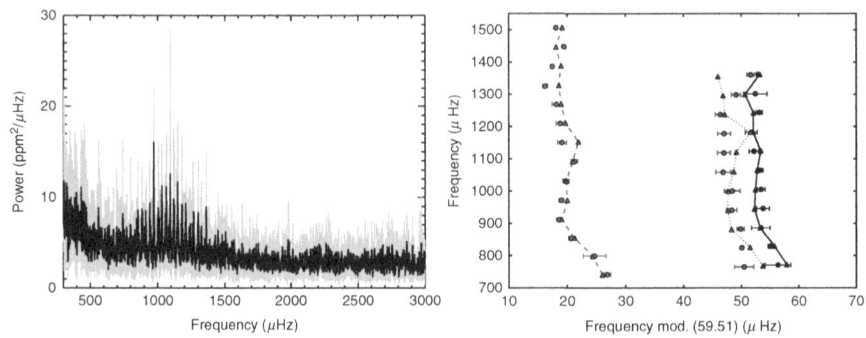

Fig. 2 *Left*: Power spectrum of HAT-P-7. *Right*: Échelle diagram for frequencies of degree $\ell = 0$ - 2 in HAT-P-7, where a large separation of 59.51 μHz is used. The *filled circles* with error bars give the observed frequencies while the *filled triangles* give the frequencies for the best TDC model. The colors indicate the spherical degree ℓ: *black* ($\ell = 0$), *blue* ($\ell = 2$) and *red* ($\ell = 1$)

3 Results

The right panel of Fig. 2 gives the échelle diagram of HAT-P-7. The diagram is built using a large separation of 59.51 μHz for frequencies of degree $\ell = 0$–2. The observed (filled circles) and theoretical frequencies (filled triangles) correspond to three identified spherical degrees illustrated by different colors: black ($\ell = 0$), blue ($\ell = 2$) and red ($\ell = 1$). The agreement between model and observations is very good.

The first four rows in Table 1 give the global parameters of the star. Our results are given in column four. They are all within the photometric observational boxes. The obtained mass and radius values lie between the isochrones fit values of [8] and the adiabatic asteroseismic values of [4]. TDC models estimation of the age agrees more with the isochrones fitting [8] than the adiabatic calculations of [4].

Table 1 Stellar and planetary parameters

	HATNet project	*Kepler* data: Q0-Q1	*Kepler* data: Q0-Q2
$M_\star(M_\odot)$	1.47[a] ± 0.08	1.52[b] ± 0.036	1.4150[d] ± 0.020
$R_\star(R_\odot)$	1.84[a] ± 0.23	1.991[b] ± 0.018	1.9276[d] ± 0.010
$T_{eff}(K)$	6350[a] ± 80	–	6423[d] ± 50
Age(Gy)	2.2[a] ± 1.0	2.14[b] ± 0.26	2.2082[d] ± 0.04
$R_P(R_J)$	1.363[a] ± 0.195	1.50[c] ± 0.02	1.4588[d] ± 0.007
$M_P(M_J)$	1.776[a] ± 0.077	1.82[c] ± 0.03	1.709[d] ± 0.024
$T_{dayside}(K)$	2730[a] ± 150	2885[c] ± 100	2811[d] ± 22

[a][8]: by fitting isochrones
[b][4]: Adiabatic asteroseismic study
[c][9]
[d]This work.

We use then these results as input to characterize the planet. The results are shown in the bottom part of Table 1. We use for that the methods described in [8] for the estimation of the dayside temperature and [10] for the other parameters. The error bars associated to our estimations are much smaller compared to the others.

In summary, our physically more robust TDC modeling of the oscillations of HAT-P-7 yields more reliable global parameters for the star. Which gives tighter constraints on the properties of its orbiting planet. The analysis of the remaining data obtained by *Kepler* is expected to give more precise frequency list. We plan to run more models in order to build a denser grid. This will allow us to explore a larger parameter space for a better characterization of the star and hence, its planet.

Acknowledgements Part of this work was supported by *FCT-MCTES-Portugal* and *FEDER-EC*, through grants PTDC/CTE-AST/098754/2008, SFRH/BPD/41270/2007 and SFRH/BPD/47611/2008.

References

1. Appourchaux, T., Gizon, L., Rabello-Soares, M.-C.: A&A, **132**, 107–119 (1998)
2. Bakos, G.Á., Torres, G., Pál, A., et al.: ApJ **710**, 1724 (2010)
3. Benomar, O., Baudin, F., Chaplin, W.J., Elsworth, Y., Appourchaux, T.: MNRAS **420**, 2178 (2012)
4. Christensen-Dalsgaard, J., Kjeldsen, H., Brown, T.M., et al.: ApJL **713**, 164 (2010)
5. Koch, D.G., Borucki, W.J., Basri, G., et al.: ApJL **713**, 79 (2010)
6. Mandel, K., Agol, E.: ApJL **580**, 171 (2002)
7. O'Donovan, F.T., Charbonneau, D., Mandushev, G., et al.: ApJL **651**, 61 (2006)
8. Pál, A., Bakos, G.Á., Torres, G., et al.: ApJ **680**, 1450 (2008)
9. Welsh, W.F., Orosz, J.A., Seager, S., et al.: ApJL **713**, 145 (2010)
10. Winn, J.N.: (2010) arXiv:1001.2010

The New Kepler Picture of Variability Among A and F Type Stars

K. Uytterhoeven and KASC WG#10

Abstract The *Kepler* spacecraft is providing photometric time series with micro-magnitude precision for thousands of variable stars, what opens up opportunities to study the pulsational variability in much more detail than was previously possible from the ground. We present a first general characterization of the variability of A-F type stars as observed in the *Kepler* light curves of a sample of 750 candidate A-F type stars.

1 Motivation

The classes of δ Scuti (δ Sct) and γ Doradus (γ Dor) stars lie next to each other along the main sequence (MS) in the Hertzsprung-Russell diagram (HR-diagram), with a small overlap of their instability strips (IS). The two classes have different pulsational properties (see the description below). It remains unclear if there exists a relation or link between them. Key objects for such an investigation are hybrid δ Sct / γ Dor stars. Based on solely ground-based observations only a couple of hybrid stars have been discovered, e.g. [5, 7]. Thanks to the long-term monitoring with photometric precision at the level of micro-magnitudes of space missions such as MOST, CoRoT, and *Kepler*, many more hybrid A-F type pulsators are being discovered [2]. In particular, the *Kepler* mission is providing high-quality time series

K. Uytterhoeven (✉)
Instituto de Astrofísica de Canarias (IAC), Tenerife
Department Astrofísica, Universidad de La Laguna (ULL), Tenerife
Laboratoire AIM, CEA/DSM-CNRS-Université Paris Diderot; CEA, IRFU, SAp, Centre de Saclay
Kiepenheuer-Institut für Sonnenphysik, Freiburg im Breisgau
e-mail: katrien@iac.es

KASC WG#10
http://astro.phys.au.dk/KASC/

Uytterhoeven, K. and KASC WG#10.: *The New Kepler Picture of Variability Among A and F Type Stars*. Astrophys Space Sci Proc. **31**, 231–235 (2013)
DOI 10.1007/978-3-642-29630-7_43,

for several hundreds of A-F type stars, allowing for the first time a comprehensive analysis of the variability of a large sample of A-F type stars.

δ **Sct stars:** Stars that pulsate in low-order gravity (g) and pressure (p) modes with periods between 15 min and 5 h (frequencies $>5\,d^{-1}$). Modes are excited through the κ-mechanism. Typical global parameters are: $\log g = 3.2$–4.3 and $T_{\rm eff} = 6{,}300$–$8{,}600$ K [6].
γ **Dor stars:** Stars that pulsate in high-order g modes with typical periods between 8 h and 3 d (frequencies $<5\,d^{-1}$). Modes are excited by a flux modulation mechanism induced by the upper convective layer [3]. Typical global parameters are: $\log g = 3.9$–4.3 and $T_{\rm eff} = 6{,}900$–$7{,}500$ K [4].

We describe here the main results of an analysis of *Kepler* survey phase data (May 2009–March 2010) of a sample of 750 candidate A-F type stars, as presented by [8]. The sample consists of stars initially assigned to KASC[1] Working Groups on γ Dor and δ Sct stars, to which stars that clearly show δ Sct or γ Dor type pulsations were added.

2 Observational Classification

We investigated the light curves, the extracted frequency spectra, and the detected frequencies of the 750 sample stars, with the aim to classify candidate γ Dor, δ Sct, and hybrid stars. We used an automatic procedure to identify obvious harmonic and alias frequencies, which were subsequently discarded. We stress that only seemingly "independent" frequencies with amplitudes above 20 ppm were considered in the analysis described below. We compiled a database of available values of physical parameters such as $T_{\rm eff}$ and $\log g$. For most stars the only available source was the Kepler Input Catalogue [1]. For about 110 stars independent $T_{\rm eff}$ and $\log g$ values were available in the literature or were derived from new ground-based data.

We classified the A-F type stars in three groups: δ Sct stars, γ Dor stars, and hybrid stars. The classification criteria used were: a δ Sct star shows mainly frequencies $> 5\,d^{-1}$; a γ Dor star shows mainly frequencies $< 5\,d^{-1}$; a star was considered a candidate hybrid star when at least two independent frequencies with amplitudes above 100 ppm were detected in both domains, whereby the highest amplitudes in the two domains are comparable or do not differ more than a factor of 5–7. We tried to avoid misclassifications due to long-term signals caused by rotation or short-term signals introduced by the spectral window. Using these criteria, we classified 27% of the sample stars as candidate δ Sct star, 13% as candidate γ Dor

[1]Kepler Asteroseismic Science Consortium, http://astro.phys.au.dk/KASC

star, and 23% (171 stars) as candidate hybrid star. *These numbers imply that hybrid δ Sct / γ Dor pulsators are not uncommon.*

The remaining 275 stars in our sample were identified as rotationally modulated/active stars, binaries, stars of different spectral type, or stars that show no clear periodic variability.

3 Observational Results

We investigated physical properties, such as $T_{\rm eff}$ and $\log g$, of the 475 *Kepler* stars classified as δ Sct, γ Dor, or hybrid star, and characterized them in terms of number of independent frequencies, amplitude heights, and range of detected frequencies.

The right panel of Fig. 1 shows the position of the 475 *Kepler* stars in the ($T_{\rm eff}$, $\log g$)-diagram, together with the observational IS for δ Sct and γ Dor stars [4, 6]. Squares, asterisks and bullets represent δ Sct, γ Dor, and hybrid stars, respectively. As a comparison, the left panel of Fig. 1 shows the δ Sct, γ Dor, and hybrid stars known from before the space mission and detected from ground-based observations.

Based on the currently available values of $T_{\rm eff}$ and $\log g$, and taking into account their (large) error bars, we find that:

- γ Dor pulsations are detected outside the γ Dor IS, in both cooler and hotter stars than previously observed from the ground.
- δ Sct pulsators are detected beyond the red edge of the δ Sct IS.
- hybrid stars are not confined to the small overlapping region of the δ Sct and γ Dor IS, and are detected in the entire region between the blue edge of the δ Sct IS and the red edge of the γ Dor IS, and beyond.

These results seem to suggest that the edges of the so far accepted observational IS need to be revised. However, we need accurate values of $T_{\rm eff}$ and $\log g$ for all stars to confirm this finding. Also, the known driving mechanisms do not explain the observed hybrid pulsations in hot and cool A-F type stars. *Alternative pulsation mechanisms need to be investigated to supplement the κ mechanism and convective blocking effect to drive hybrid pulsations.*

Another finding is that there is a great concentration of stars near the overlap of the δ Sct and γ Dor IS: *hybrid, δ Sct, and γ Dor stars coincide in the same region of the HR-diagram.* The fact that stars with similar temperatures show a different pulsational behaviour raises questions, e.g., What makes a star a δ Sct, γ Dor, or hybrid pulsator? What is the difference in their internal physics?

When investigating the number of independent frequencies, the associated amplitudes, and the observed frequency range for the three groups of A-F type pulsators, we find:

- Up to 500 non-combination frequencies are detected in the variability of the *Kepler* light curves of A-F type pulsators. However, the majority (66%) shows less than 100 independent variable signals.

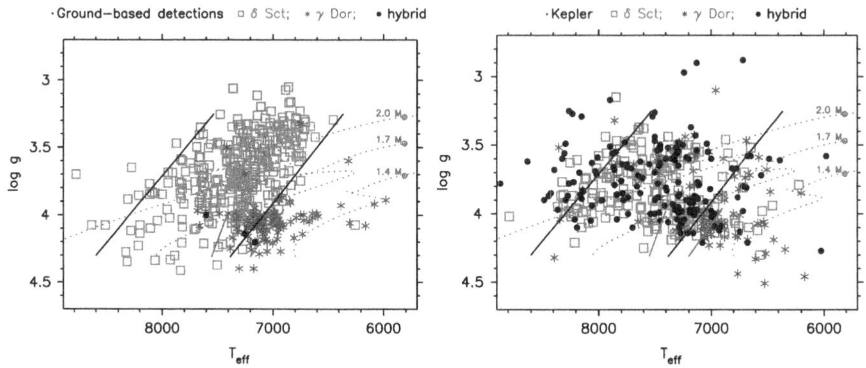

Fig. 1 (*left*): ($T_{\rm eff}$, $\log g$)-diagram of δ Sct, γ Dor, and hybrid stars detected from the ground. (*right*): ($T_{\rm eff}$, $\log g$)-diagram of the 475 *Kepler* stars classified as A-F type pulsators. The *solid thick black and light grey lines* mark the *blue and red edge* of the observed IS of δ Sct and γ Dor stars [4, 6]. The *dotted lines* mark evolutionary tracks for MS stars with different masses

- The number of detected frequencies versus $T_{\rm eff}$ follows a distribution that peaks between the centre and the red edge of the respective IS.
- For about 59% of the stars the highest amplitude is lower than 2,000 ppm. In general, higher amplitudes are detected in δ Sct stars than in γ Dor stars.
- The "frequency gap" between 5–10 $\rm d^{-1}$ predicted by the current instability models for hybrid stars is not observed in the *Kepler* data.

To summarize, *hybrid δ Sct/γ Dor pulsations seem to be more common than previously anticipated* and are found in stars with a large range of temperatures. These results open up new questions and demand a further investigation of the pulsational driving mechanisms in A-F type stars. Thanks to the high-quality *Kepler* data of hundreds of A-F type stars that are becoming available we have good prospects in making progress in this field.

Acknowledgements KU acknowledges financial support by the Deutsche Forschungsgemeinschaft (DFG) in the framework of project UY 52/1-1, and by the Spanish National Plan of R&D for 2010, project AYA2010-17803.

References

1. Brown, T.M., Latham, D.W., Everett, M.E., Esquerdo, G.A.: Kepler input catalog: photometric calibration and stellar classification. AJ **142**, 112 (2011)
2. Grigahcène, A., Antoci, V., Balona, L., et al.: Hybrid γ Doradus-δ Scuti pulsators: New insights into the physics of the oscillations from *Kepler* observations. ApJ **713**, L192 (2010)
3. Guzik, J.A., Kaye, A.B., Bradley, P.A., Cox, A.N., Neuforge, C.: Driving the gravity-mode pulsations in γ Doradus variables. AJ **542**, 57 (2000)

4. Handler, G., Shobbrook, R.R.: On the relationship between the δ Scuti and γ Doradus pulsators. MNRAS **333**, 251 (2002)
5. Henry, G.W., Fekel, F.C.: HD 8801: A unique single Am star with γ Doradus and δ Scuti pulsations. AJ **129**, 2026 (2005)
6. Rodríguez, E., Breger, M.: Delta Scuti and related stars: Analysis of the R00 catalogue. A&A **366**, 178 (2001)
7. Uytterhoeven, K., Mathias, P., Poretti, E., et al.: The γ Doradus CoRoT target HD 49434. I. Results from the ground-based campaign. A&A **489**, 1213 (2008)
8. Uytterhoeven, K., Moya, A., Grigacène, A., et al.: The *Kepler* characterization of the variability among A- and F-type stars. I. General overview. A&A (2011) doi: 10.1051/0004-6361/201117368

Stellar Activity Cycles and Contribution of the Deep Layers Knowledge

S. Mathur

Abstract It is believed that magnetic activity on the Sun and solar-type stars are tightly related to the dynamo process driven by the interaction between rotation, convection, and magnetic field. However, the detailed mechanisms of this process are still incompletely understood. Many questions remain unanswered, e.g.: why some stars are more active than others?; why some stars have a flat activity?; why is there a Maunder minimum?; are all the cycles regular? A large number of proxies are typically used to study the magnetic activity of stars as we cannot resolve stellar discs. Recently, it was shown that asteroseismology can also be used to study stellar activity, making it an even more powerful tool. If short cycles are not so uncommon, we expect to detect many of them with missions such as CoRoT, *Kepler*, and possibly the PLATO mission. We will review some of the latest results obtained with spectroscopic measurements. We will show how asteroseismology can help us to better understand the complex process of dynamo and illustrate how the CoRoT and *Kepler* missions are revolutionizing our knowledge on stellar activity. A new window is being opened over our understanding of the magnetic variability of stars.

1 Why Study Stellar Magnetic Activity?

The study of magnetic cycles in stars other than the Sun is important from the point of view of *solar-stellar connexion*. The detailed mechanisms of the dynamo are not yet well known as we still have difficulties in predicting the solar cycles. An example is the solar minimum between cycles 23 and 24 that was unexpectedly long, e.g. [26, 46]. We know that the dynamo results from the interaction between

---

S. Mathur (✉)
High Altitude Observatory, NCAR, P.O. Box 3000, Boulder, CO 80307, USA
e-mail: savita@ucar.edu

Mathur, S.: *Stellar Activity Cycles and Contribution of the Deep Layers Knowledge*.
Astrophys Space Sci Proc. **31**, 237–243 (2013)
DOI 10.1007/978-3-642-29630-7_44,

rotation (latitudinal differential rotation), convection, and magnetic field. By studying stellar cycles, we can expect to better understand how dynamos operate under several physical conditions (rotation rate, depth of convective zone), assuming that dynamo operates similarly in solar-type stars compared to the Sun. This work would put additional constraints on solar dynamos models, for which, there is a variety of scenarios: the mean-field model, the thin-shell model, and the distributed dynamos, e.g. [18, 19, 28, 32]. As the dynamo takes place in the outer layers of the stars, it is required to have a good knowledge on their structure. For instance, the position of the base of the convective zone is a crucial ingredient and it seems that the tachocline plays a role in the efficiency of the dynamo but its exact contribution remains unclear. Over the past decade, asteroseismology has emerged as a powerful unique tool allowing us to directly probe the stellar interior. Aside from determining the radii, masses, and ages of stars more precisely than with classical techniques [37], with asteroseismology we can fit stellar models with observed frequencies [35] and, for example, use acoustic glitches [4] to have estimates of the depth of the convection zone of the stars. Beside, using photometric observations, different techniques have been developed or applied to study the surface rotation, such as spot modeling, interferometric imaging or wavelet techniques. The detection of mixed modes [8, 10, 14, 34] can also provide information on the rotation of the stellar interior, specially the core, as these modes are the result of the coupling between acoustic (p) and gravity (g) modes. The NASA mission *Kepler* [12] has already made a step further concerning this topic by unveiling the radial differential rotation in a few red-giant stars [9]. By combining this information on the rotation profile of a star and its magnetic activity, we should have more constraints on the stellar dynamo models, e.g. [13].

Another impact of the stellar activity study would be in the field of the *star-planet interaction*. Within a planetary system, stars are the largest source of energy [45]. The emission of particles, through winds for instance, can affect the composition, thermal properties and the atmosphere of the planet(s). The existence and the measurement of the strength of stellar winds were studied with H Lyman α lines [52]. These winds can be responsible for the increase of atmospheric temperature of the planet and erosions on its surface. These particle emissions can also trigger some photochemical reactions [6]. Observations of X-rays and far UV (FUV) for several stars with different ages were analyzed by [44] who showed that the older the star gets, the lower the X-ray and FUV fluxes become. These radiations are important to study the development of life on the planets and bring another factor to take into account for the definition of the habitable zones of exoplanets. Finally, there is a debate about the impact of the planet on the activity of host stars. Some tidal and magnetic scenarios have been developed to explain these phenomena while [49], for example, suggested that planets could trigger a release of energy that was built up in coronal loops. Several teams also tried to study if close-in giant planets can have an impact on host-stars X-ray emission but so far no clear signature has been found [27, 42, 48].

2 Classical Observations

A large number of proxies are typically used to study the magnetic activity of stars, such as optical, UV or X-ray emission from magnetically heated gas but also starspots or flares, and other phenomena that are associated with magnetic fields in the Sun. Many surveys have been carried out to measure UV, X-ray and CaHK in the stellar atmosphere. The common point between all of them is that they depend on the surface activity and possibly on the inclination angle of the star and the position of the active latitudes. In this section, we will present the results that have been obtained on stellar cycles thanks to these proxies.

2.1 The Mount Wilson Project and Some Results

One of the most used proxy is the Mount Wilson S-index, which is sensitive to the photosphere and the chromosphere. It requires a high contrast between the atmospheric energy and the underlying thermal spectrum, making ground-based observations possible for these kinds of observations. It is thus defined as the ratio between the flux in HK lines and the flux in R and V bandpasses [51]. The average value for the Sun is around 0.171. Another proxy, $\log(R'_{\rm HK})$, is sensitive to the chromosphere. It is defined as the ratio between the emission from chromosphere in CaHK and the total bolometric emission of the star. The typical average value for the Sun is of -4.9.

In the 1960s a huge project of CaHK observations started to study the temporal variation of chromospheric emission in a large amount of main-sequence stars. This project was led at the Mount Wilson Observatory, leading to 40 years of data.

From a sample of $\sim$110 stars monitored by the project, [3] showed that the activity level, the rotation period, the presence of a Maunder minimum, the regularity of the cycles could vary with the evolutionary stage of the star. They also found that 60% of lower MS stars are similar to the Sun, 25% of stars seem to be variable, and 15% of the stars seem to have a flat activity.

By analyzing another sample of stars monitored by this project, [43] suggested that the total stellar irradiance varies with chromospheric activity. In particular, young and more active stars appear to be fainter when HK emission increases while old and less active stars (like Sun) are brighter when HK emission increases. This is in agreement with what we observe with the Sun where the brightness increases when more sunspots are present on its surface.

Some studies were also done to find a correlation between the cycle period and the rotation period of the star. For cool stars like the Sun, with an $\alpha\Omega$ dynamo, the longer the period of rotation is, the longer the cycle period will be [50]. Indeed, an empirical formula has been derived (e.g. [29]) where $P_{\rm cyc}/P_{\rm rot}$ depends on the convective turnover time. But due to large uncertainties on some structural parameters (e.g. depth of the convective zone), it is difficult to have a precise estimate of P_{cyc}. Finally, [11] studied some of the Wilson project stars and found a very interesting result showing that two different branches seem to exist among the stars: the active and inactive branches. No clear explanation has been given so far but some possible scenarios involve the α effect or maybe different locations of the dynamo shell. The position of the Sun is also striking as it is just between the two branches.

2.2 Some Interesting Cases

A solar twin, 18 Sco: This solar twin has a mass and a radius very close to the ones of the Sun [7]. Petit et al. [40] estimated its rotation period to 22.7±0.5 days. Using $\sim$ 10 years of spectroscopic and photometric data from the Solar-Stellar Spectrograph and the Automatic Photospheric Telescope, an excess flux was found in the CaHK measurements [25]. Beside, simultaneous measurements in b and y passbands were done and they presented similar variations to the Sun. The average Mount Wilson S-index was estimated to $\sim$ 0.182 and the cycle period $P_{cyc} \sim 7$ years.

Short activity cycles: The star ι Hor (or HD170151) is known to be hosting an exoplanet. It has been observed by the Cerro Tololo Observatory in Chile for 2.5 years. The Mount Wilson S-index of these data varied from 0.21 to 0.28, which is around 30% more than what we see in the Sun. The Lomb–Scargle periodogram suggests a rotation period of 8 days, which means that this is a rather fast rotator. Metcalfe et al. [38] fitted the data and found a cycle period of 1.6 years, putting this star in the category of stars with the shortest cycle periods detected. Another very short cycle was observed for the planet hosting star, τ Boo, for which two magnetic reversals have been observed [22], suggesting a cycle period of 1 year. These results are very promising as it suggests that short cycles are more common than we thought and with the present instrumentation, we could expect to observe some complete magnetic activity cycles in other stars.

Magnetic field reversal: Today, it is possible to study the magnetic field topology of stars using spectropolarimetric measurements with instruments such as ESPaDOnS [20] or NARVAL[1]. Petit et al. [41] studied the evolution of the large-scale photospheric magnetic field geometry of the solar-type star HD 190771 with NARVAL data. This star is a solar analog as well but with a rotation period of $\sim$ 8 days. Their analysis put in evidence a polarity reversal between observations separated by 1 year as well as dramatic change of the toroidal field over a period of a few months. This kind of information is very important to understand the impact of the rotation on the generation of magnetic fields through the dynamo mechanism.

3 A New Era with Seismology

The new generation of space missions such as CoRoT [2] and *Kepler* are allowing us to do asteroseismology by providing exquisite photometric data [17]. Among the solar-type stars, many light curves present some variability that can be attributed to the presence of starspots crossing the stellar surface and thus suggesting that these stars are magnetically active. One way of studying the magnetic activity of these stars (apart from using classical observations as in Sect. 2), we can use spot modeling [21, 30], which has already given results with some CoRoT and *Kepler* targets [5, 23, 31, 33, 39]. This modeling is based on a large number of parameters but can retrieve estimates of the lifetime of the spots, the surface rotation, and the presence of differential rotation.

Beside, we also know that for the Sun, when the magnetic activity increases, the frequency of the p modes increases while their amplitude decreases [46]. The reason why we observe these changes are still under investigation but they are probably related to changes in the outer layers of the stars that affect the outer turner point of the modes. Aside that, it also seems that though no surface magnetic activity is observed (such as a Maunder minimum) a possible cycle could still be running underneath the photosphere. Prediction of the amplitude of frequency shifts using two different scalings between the frequency shift and the variation of the R'_{HK} index have been done. While [15] predicted higher amplitudes for G stars, [36] predicted higher amplitudes for F stars. Recently, [24] discovered the first signature of magnetic activity using seismology in an F star: HD 49933, agreeing with the predictions by [36]. CaHK observations in Chile confirmed that this star is magnetically active with a cycle period of at least 210 days. A deeper analysis of the frequency shifts of the p modes as a function of the frequency reveals that the high-frequency modes are more shifted than the low-frequency modes, which is similar to what we observe in the Sun [47], suggesting that these frequency shifts are related to a perturbation produced by magnetic activity occurring in the outer layers of the star. A similar analysis has been done for three other CoRoT targets [34] that were also observed by the spectropolarimeter NARVAL. There is some hint of activity but more data would be necessary to confirm any clear detection. An ensemble analysis of hundreds of solar-type stars observed by *Kepler* also showed that stars that seem to be more active have lower p-mode amplitude and even undetectable modes [16].

We are starting to have many observations of magnetic activity and cycles in other stars while asteroseismology is also providing information on the stellar interiors. It is just a question of time to combine both information to improve our understanding of dynamo process and how the magnetic field evolves in the outer layers of the stars.

Acknowledgements NCAR is supported by the National Science Foundation. This research was supported in part by the National Science Foundation under Grant No. NSF PHY05-51164.

References

1. Aurière, M.: Stellar Polarimetry with NARVAL ESA Publications Series, vol. 9, p. 105 (2003)
2. Baglin, A., Auvergne, M., Boisnard, L., et al.: CoRoT: a high precision photometer for stellar evolution and exoplanet finding. COSPAR, Plenary Meeting **36**, 3759 (2006)
3. Baliunas, S.L., Donahue, R.A., Soon, W.H., et al.: Chromospheric variations in main-sequence stars. ApJ **438**, 269–287 (1995)
4. Ballot, J., Turck-Chièze, S., García, R.A.: Seismic extraction of the convective extent in solar-like stars. The observational point of view. A&A **423**, 1051–1061 (2004)
5. Ballot, J., Gizon, L., Samadi, R., et al.: Accurate p-mode measurements of the G0V metal-rich CoRoT target HD 52265. A&A **530**, 97–108 (2011)
6. Bauer, S.J., Lammer, H.: Planetary aeronomy: atmosphere environments in planetary systems. Springer (2004)
7. Bazot, M., Ireland, M.J., Huber, D., et al.: The radius and mass of the close solar twin 18 Scorpii derived from asteroseismology and interferometry. A&A **526**, L4 (2011)
8. Beck, P.G., Bedding, T.R., Mosser, B., et al.: *Kepler* detected gravity-mode period spacings in a red giant star. Science **332**, 205 (2011)
9. Beck, P.G., Montalban, J., Kallinger, T., et al.: Fast core rotation in red giant stars as revealed by gravity-dominated mixed modes. Nature **481**, 55 (2012)
10. Bedding, T.R., Mosser, B., Huber, D., et al.: Gravity modes as a way to distinguish between hydrogen- and helium-burning red giant stars. Nature **471**, 608–611 (2011)
11. Böhm-Vitense, E.: Chromospheric activity in G and K main-sequence stars, and what it tells us about stellar dynamos. ApJ **657**, 486–493 (2007)
12. Borucki, W.J., Koch, D., Basri, G., et al.: *Kepler* planet-detection mission: introduction and first results. Science **327**, 977–980 (2010)
13. Brown, T.M., Latham, D.W., Everett, M.E., Esquerdo, G.A.: *Kepler* input catalog: photometric calibration and stellar classification. AJ **142**, 112–130 (2011)
14. Campante, T.L., Handberg, R., Mathur, S., et al.: Asteroseismology from multi-month *Kepler* photometry: the evolved Sun-like stars KIC 10273246 and KIC 10920273. A&A **534**, 6–20 (2011)
15. Chaplin, W.J., Houdek, G., Appourchaux, T., et al.: Challenges for asteroseismic analysis of Sun-like stars. A&A **485**, 813–822 (2008)
16. Chaplin, W.J., Bedding, T.R., Bonanno, A., et al.: Evidence for the impact of stellar activity on the detectability of solar-like oscillations observed by *Kepler*. ApJ **732**, L5–L11 (2011)
17. Chaplin, W.J., Kjeldsen, H., Christensen-Dalsgaard, J., et al.: Ensemble asteroseismology of solar-type stars with the NASA *Kepler Mission*. Science **332**, 213–216 (2011)
18. Choudhuri, A.R., Schussler, M., Dikpati, M.: The solar dynamo with meridional circulation. A&A **303**, 29–32 (1995)
19. Dikpati, M.: Solar magnetic fields and the dynamo theory. Adv. Spa. Res. **35**, 322–328 (2005)
20. Donati, J.-F.: ESPaDOnS: An Echelle SpectroPolarimetric Device for the Observation of Stars at CFHT. ASPC Series **307**, 41 (2003)
21. Dorren, J.D.: A new formulation of the starspot model, and the consequences of starspot structure. ApJ **320**, 756–767 (1987)
22. Fares, R., Donati, J.-F., Moutou, C., et al.: Magnetic cycles of the planet-hosting star τ Bootis - II. A second magnetic polarity reversal. MNRAS **398**, 1383–1391 (2009)
23. Frasca, A., Frhlich, H.-E., Bonanno, A., et al.: Magnetic activity and differential rotation in the very young star KIC 8429280. A&A **532**, 81 (2011)
24. García, R.A., Mathur, S., Salabert, D., et al.: CoRoT reveals a magnetic activity cycle in a Sun-like star. Science **329**, 1032 (2010)
25. Hall, J.C., Lockwood, G.W.: Evidence of a pronounced activity cycle in the solar twin 18 scorpii. ApJ **545**, L43–L45 (2000)
26. Howe, R., Christensen-Dalsgaard, J., Hill, F., et al. A note on the torsional oscillation at solar minimum. ApJL **701**, 87–90 (2009)

27. Kashyap, V.L., Drake, J.J., Saar, S.H.: Extrasolar giant planets and x-ray activity. ApJ **687**, 1339–1354 (2008)
28. Jouve, L., Brun, A.S., Arlt, R., et al.: A solar mean field dynamo benchmark. A&A **483**, 949–960 (2008)
29. Jouve, L., Brown, B.P., Brun, A.S.: Exploring the P_{cyc} vs. P_{rot} relation with flux transport dynamo models of solar-like stars. A&A **509**, 32 (2010)
30. Lanza, A.F., Rodono, M., Pagano, I., et al.: Modelling the rotational modulation of the Sun as a star. A&A **403**, 1135–1149 (2003)
31. Lanza, A.F., Aigrain, S., Messina, S., et al.: Photospheric activity and rotation of the planet-hosting star CoRoT-4a. A&A **506**, 255–262 (2009)
32. MacGregor, K.B., Charbonneau, P.: Solar interface dynamos. I. Linear, kinematic models in cartesian geometry. ApJ **486**, 484–501 (1997)
33. Mathur, S., García, R.A., Catala, C., et al.: The solar-like CoRoT target HD 170987: spectroscopic and seismic observations. A&A, **518**, 53–65 (2010)
34. Mathur, S., Handberg, R., Campante, T.L., et al.: Solar-like Oscillations in KIC 11395018 and KIC 11234888 from 8 Months of *Kepler* data. ApJ **733**, 95–105 (2011)
35. Mathur, S., Metcalfe, T., Woitaszek, M., et al.: A uniform asteroseismic analysis of 22 solar-type stars observed by *Kepler*. ApJ **749**, 152 (2012)
36. Metcalfe, T.S., Dziembowski, W.A., Judge, P.G., et al.: Asteroseismic signatures of stellar magnetic activity cycles. MNRAS **379**, L16–L20 (2007)
37. Metcalfe, T.S., Monteiro, M.J.P.F.G., Thompson, M.J., et al.: A precise asteroseismic age and radius for the evolved Sun-like star KIC 11026764. ApJ **723**, 1583–1598 (2010)
38. Metcalfe, T.S., Basu, S., Henry, T.J., et al.: Discovery of a 1.6 year magnetic activity cycle in the exoplanet host star ι Horologii. ApJ **723**, L213–L217 (2010)
39. Mosser, B., Baudin, F., Lanza, A.F., et al.: Short-lived spots in solar-like stars as observed by CoRoT. A&A **506**, 245–254 (2009)
40. Petit, P., Dintrans, B., Solanki, S.K., et al.: Toroidal versus poloidal magnetic fields in Sun-like stars: a rotation threshold. MNRAS **388**, 80–88 (2008)
41. Petit, P., Dintrans, B., Morgenthaler, A., et al.: A polarity reversal in the large-scale magnetic field of the rapidly rotating sun HD 190771. A&A **508**, L9–L12 (2009)
42. Poppenhaeger, K., Schmitt, J.H.M.M.: A correlation between host star activity and planet mass for close-in extrasolar planets? ApJ **735**, 59 (2011)
43. Radick, R.R., Lockwood, G.W., Skiff, B.A., et al.: Patterns of variation among Sun-like stars. ApJS **118**, 239–258 (1998)
44. Ribas, I., Guinan, E.F., Güdel, M., et al.: Evolution of the solar activity over time and effects on planetary atmospheres. I. High-energy irradiances (1–1700). ApJ **622**, 680–694 (2005)
45. Ribas, I., Porto de Mello, G.F., Ferreira, L.D., et al.: Evolution of the solar activity over time and effects on planetary atmospheres. II. κ^1 Ceti, an analog of the Sun when life arose on Earth. ApJ **714**, 384–395 (2010)
46. Salabert, D., García, R.A., Pallé, P.L., et al.: The onset of solar cycle 24. What global acoustic modes are telling us. A&A **504**, L1–L4 (2009)
47. Salabert, D., Régulo, C., Ballot, J., et al.: About the p-mode frequency shifts in HD 49933. A&A **530**, 127 (2011)
48. Scharf, C.A.: Possible constraints on exoplanet magnetic field strengths from planet-star interaction. ApJ **722**, 1547–1555 (2010)
49. Shkolnik, E., Walker, G.A.H., Bohlender, D.A., et al.: Hot Jupiters and Hot spots: The short- and long-term chromospheric activity on stars with giant planets. ApJ **622**, 1075–1090 (2005)
50. Saar, S.H., Brandenburg, A.: A new look at dynamo cycle amplitudes. AN **323**, 357–360 (2002)
51. Wilson, O.C.: Chromospheric variations in main-sequence stars. ApJ **226**, 379–396 (1978)
52. Wood, B.E., Müller, H.-R., Zank, G.P., et al.: Measured mass-loss rates of solar-like stars as a function of age and activity. **574**, 412–425 (2002)

Part VI
Early Type Stars. Oscillations Versus Other Agents, Mainly Regular/Rotational Variations of B Stars

Rotational Modulation in Be Stars

L.A. Balona

Abstract The broad peaks in the periodograms of the light variations of Be stars do not resolve into individual frequencies even in a long time series. Recent *CoRoT* observations, in particular, clearly demonstrate their stochastic nature. These peaks, which have frequencies indistinguishable from the rotational frequency and its first harmonic, have generally been attributed to $l = 2$ nonradial pulsation (NRP). Stochastic variation with the rotational period is an inevitable consequence of localized mass loss for any mass-loss mechanism, including NRP. It is evident that the broad peaks are just a result of co-rotating circumstellar gas clouds. A simple model in which the mass loss is a result of magnetic flaring explains all major characteristics of Be stars.

1 Introduction

Be stars are dwarf and giant B stars in which the Balmer lines, particularly Hα, show emission or incipient emission at some time. The emission arises from a circumstellar gas disc in Keplerian rotation [9]. The origin of the circumstellar gas is to be found in episodic outbursts resulting in ejection of material from the photosphere. Be stars generally have broad spectral lines which implies rapid rotation. Figure 1 shows the mean equatorial velocities as a function of spectral type for Be stars obtained from the projected rotational velocities by deconvolution assuming random orientation of the axes of rotation. For early-type Be stars, the mean equatorial velocity, v, falls well below the estimated critical rotational velocity, $v_{\rm crit}$, such that $v/v_{\rm crit} \approx 0.6$. This ratio increases steadily towards cooler

L.A. Balona (✉)
South African Astronomical Observatory, P.O. Box 9, Observatory 7935, Cape Town, South Africa
e-mail: lab@saao.ac.za

Balona, L.A.: *Rotational Modulation in Be Stars*.
Astrophys Space Sci Proc. **31**, 247–252 (2013)
DOI 10.1007/978-3-642-29630-7_45, © Springer-Verlag Berlin Heidelberg 2013

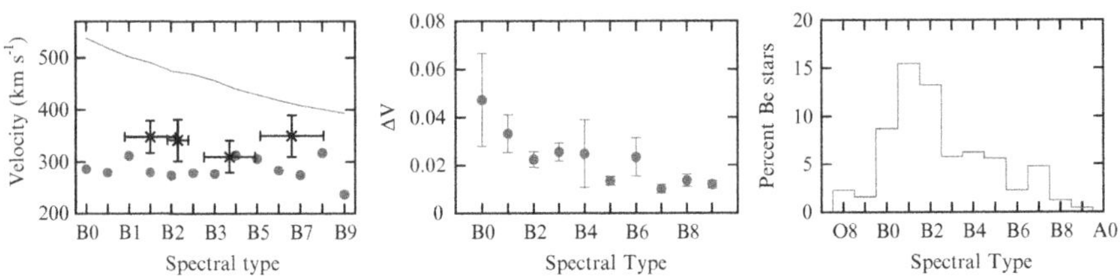

Fig. 1 *Left panel*: the mean equatorial rotational velocities of Be stars (*filled circles*) and equatorial velocities derived from the photometric periods (*crosses with error bars*) as a function of spectral type. The curve is the estimated critical rotational velocity. *Middle panel*: root mean square amplitudes in the V band as a function of spectral type for λ Eri stars. *Right panel*: the percentage of Be stars relative to all B stars of a given spectral type

stars, reaching $v/v_{\rm crit} \approx 0.8$ for late Be stars. It seems that only the late Be stars may be rotating near critical rotational velocities.

The origin of the sudden outbursts which characterize mass loss in Be stars has been linked to their rapid rotation. A popular model presumes that the ejections are triggered by pulsation. However, for pulsation to act as a trigger, the horizontal component of pulsational velocity must add to the rotational velocity so as to exceed the circular orbital velocity at the equator. Since the pulsational velocity is, at most, only some tens of km s^{-1}, the star needs to be rotating very close to the critical limit. At these very high rotational speeds, gravity darkening will lead to smaller line broadening and $v \sin i$ will be lower than the true value [14]. The pulsation model thus presumes that all Be stars are rotating very close to critical, but there is no explanation why early Be stars require a large gravity-darkening correction, while line broadening in late Be stars needs practically no adjustment for gravity darkening. In fact, the calculated correction is in the opposite sense [14].

A statistical analysis of the projected rotational velocities which takes into account the effect of gravity darkening just described, shows that the Be stars exhibit a roughly uniform spread of intrinsic rotation speed that extends from 40–60% up to 100% of critical [5]. Late-type Be stars show progressively narrower ranges of rotation speed as the effective temperature decreases. The lower limit rises to critical rotation for the coolest Be stars. The early-type Be stars were found to possess rotational velocities significantly lower than critical [5]. In other words, the pulsational model coupled with gravity darkening fails to account for the observed line broadening distribution as a function of spectral type.

The low-degree line profile variations in the large majority of early-type Be stars is consistent with non-radial pulsation (NRP). These can be modeled with spherical harmonic degree $l = m = 2$ [10]. For some reason, late-type Be stars do not seem to show pulsation at significant amplitudes, a fact which cannot be understood if NRP is central to the Be phenomenon [1]. Neither is it possible to understand why many Be stars show no signs of pulsation at all. Ground-based surveys show that 60% of early-type Be stars are periodic light variables (λ Eri stars), but this number drops to about 10% for the late-Be stars. About half of the periodic Be stars show double wave light curves [3, 6]. A star may change from a single-wave to a double-wave or vice-versa on a timescale of a few months [2]. A statistical analysis of the photometric periods and the periods derived from standard, uncorrected, $v \sin i$ measurements together with radii estimated from the spectral types, shows that the fundamental photometric period is the same as the rotation period [4].

In this note, we suggest that the periodic line profile and light variations are more easily understood as a simple consequence of a localized mass ejection process. Localized mass ejection due to any mechanism, including NRP, must give rise to gas concentrations (clouds) which initially co-rotate with the star before dissipating. This model offers a simple explanation for all major characteristics of Be stars, but does not preclude pulsation. Since Be stars fall in the β Cep and SPB instability strips, many could be pulsating. However, pulsation plays no role in the Be phenomenon and critical rotation is not a requirement for mass loss.

2 Periodic Light Variations

Space observations of Be stars confirm the general photometric behaviour seen from the ground. The *MOST* and *CoRoT* satellites photometrically detected what appear to be multiple periods in several Be stars. Periodograms of some of these stars are shown in Fig. 2. Note the characteristic broad peak and its harmonic. These variations have been interpreted as nonradial g mode pulsations of high radial order driven by the κ mechanism. Because of the low frequencies of high-order g-modes in the co-rotating frame, the observed frequency is close to the rotational frequency and its first harmonic for an $l = 2$ mode.

What has not been generally realized is that there is no strict multiperiodicity in the light curves from space or from the ground. The peaks in the periodograms of these stars remain broad, no matter how long the star is observed. A periodic signal will appear as a very narrow peak, indistinguishable from a delta function, in Fig. 2. If the observations are divided into two independent subsets, the frequencies in one subset do not match the frequencies in the other subset. The light variations are, in fact, stochastic in nature.

A direct demonstration of the stochastic nature of these variations is provided by the *CoRoT* observations of HD 49330 [7] (Fig. 2). Note the sudden appearance of the broad features at about 1.5 d^{-1} and 3.0 d^{-1} at the "precursor" stage and their dramatic decrease in intensity at the "relaxation" stage. The star is also a β Cep/SPB pulsator. Compare the two broad, temporary, features with the narrow peaks due to pulsation.

Variations in intensity and structure of the broad features are a characteristic property of Be stars and reflects the great variation in light curve shapes and amplitudes observed from the ground. Since the line profile variations and the light curves are concomitant, it follows that the line profile variations are merely a manifestation of this stochastic process and nothing to do with pulsation.

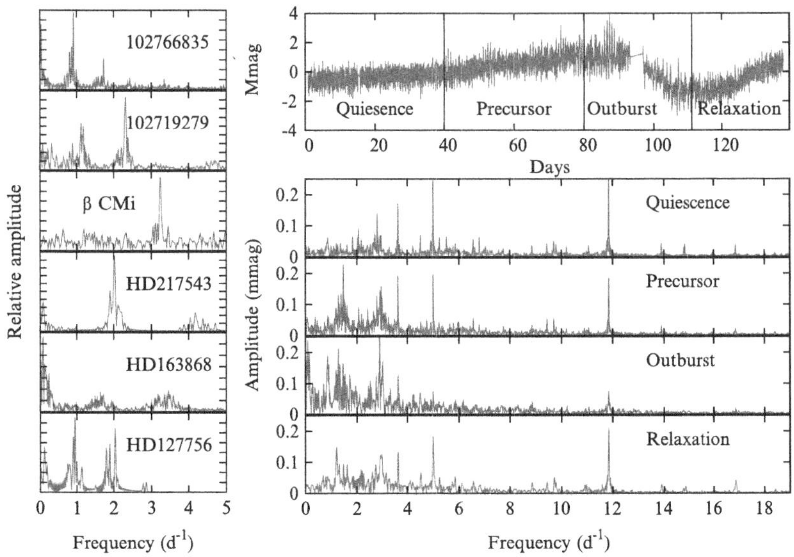

Fig. 2 *Left panel*: periodograms of some Be stars observed by *MOST* and by *CoRoT*. *Right panel*: the light curve of HD 49330 is shown at the top. The other panels show the periodograms of four sections of the light curve

Remarkably, the amplitudes of all the β Cep peaks in Fig. 2 are reduced by the same amount during the outburst. There is no known case of such rapid amplitude variations in a β Cep star. Clearly, it is not the pulsational amplitudes which have suddenly changed, but rather a sudden change in photospheric obscuration, leading to reduced visibility during the outburst. It is likely that many Be stars may be β Cep and/or SPB variables, but the pulsations are not visible due to such obscuration.

3 Rotational Velocities

The material we see in the circumstellar disc must have originated at some region in the photosphere. At the earliest and densest stage, the gas would have been localized and rotating with the star, leading to a light variation with a period equal to the period of rotation. This is an inescapable consequence of any localized mass-loss mechanism (including NRP). As the gas dissipates, phase coherence is lost, leading to the stochastic variations.

It is acknowledged, even in the NRP model, that the fundamental period is equal, or very close, to the rotation period of the star. Since the stellar radius can be estimated from its spectral type, the equatorial rotational velocity can be determined. Although the radius has a large error, the mean equatorial velocity of several stars can be obtained with a relatively small random error. The method was applied to 84 λ Eri stars between B0 and B9, and the results shown in Fig. 1. It can be seen that the equatorial velocities obtained in this way are in good agreement with those determined from $v \sin i$. This test, which is independent of gravity darkening, shows that Be stars do not rotate at critical velocities.

The two broad peaks in the periodogram are, of course, a reflection of the $l = 2$ light distribution seen in the line profile variations and the double-wave light curves. This light distribution implies that matter must be temporarily trapped at two diametrically opposite points on the photosphere. The only way that this can occur is through the action of a (weak) dipolar magnetic field. X-ray "flares" have been detected in some Be stars such as γ Cas, HD 110432 and λ Eri [12, 13]. Flaring of this nature could have a magnetic origin. Weak magnetic fields, in the range 20–150 G, have indeed been detected in eight classical Be stars [8, 15], but none of these results have been confirmed [11].

The activity in visible light can be investigated by calculating the rms light amplitude for the λ Eri stars. Figure 1 show the rms amplitude in the V band as a function of spectral type. As can be seen, the activity drops sharply for cooler stars, which is perplexing for the NRP model. Late Be stars do not show line profile variations and *CoRoT* observations indicate extremely low photometric amplitudes in late Be stars. In Fig. 1 we also show the relative numbers of Be stars at various spectral types. This distribution clearly presents a clue to the origin of the Be phenomenon, but NRP offers no explanation for this distribution.

4 The Magnetic Rotator Model

We have seen that the broad features in the periodogram are stochastic. They can be understood as a natural consequence of mass ejection. Mass loss occurs in an outburst and is not steady, as one might expect in the NRP model. This suggests a flaring event triggered by release of magnetic energy. The highly-ionized gas released in the process will expand and be trapped in closed magnetic loops. If we presume that the magnetic field configuration is an inclined dipole, gas will be trapped in a torus inclined to the rotational axis, which accounts for the double-wave light and line profile variations. Gas will escape along open field lines and will be dragged and accelerated by the rapid rotation of the star. Depending on the rotational velocity, the gas will reach circular orbital speeds at some distance from the star if it is still ionized. After some time the gas, predominantly H and He, will cool down, become neutral (and thus unaffected by the magnetic field) and slowly dissipate into the inner circumstellar disk, rotating at Keplerian velocity.

In this model, whether a circumstellar disk is formed or not depends on two factors: (a) the presence of a tilted dipole magnetic field, (b) sufficiently rapid rotation to reach circular velocity while the gas is still ionized. In the intense radiation field of a B0 star, the gas will remain ionized to large distances. Even if the star is rotating well below critical velocity, the long lever arm provided by the magnetic field will still ensure that the gas attains circular velocity while still ionized and under control of the magnetic field. On the other hand, the weak radiation field of a B9 star limits the ionization radius close to the photosphere. The lever arm is therefore short and requires a star already rotating close to critical for the gas to reach circular velocity while still ionized.

It is evident that the condition to attain circular velocity occurs in a far larger range of rotational velocities in early-type stars compared to late-type stars. It follows that early-type Be stars will have a wide range of equatorial rotational velocities. This also explains why most Be stars will be of early type. Since the co-rotating ionized region is much larger in early-type stars, it follows that activity will be higher in these stars. In this simple way, three of the most puzzling aspects of Be stars can be understood.

The detection of magnetic fields in Be stars is controversial. However, only a weak magnetic field (tens of Gauss) is required to dominate the dynamics of ionized circumstellar gas. This is well below the current detection level. It is difficult to believe that no magnetic field exists in Be stars. After all, they are rapidly rotating conducting spheres in the presence of a Galactic magnetic field.

5 Conclusion

It needs to be appreciated that Be stars are very complex. Some Be stars, no doubt, will owe their emission to binarity. It is even possible that NRP might be effective in mass loss in a few stars. Moreover, there is little doubt that many are probably (possibly hidden) β Cep and SPB variables. It is nevertheless true that NRP cannot account for most of the Be star properties, while the magnetic rotator model does so in a very simple way.

References

1. Baade, D.: A&A **222**, 200 (1989)
2. Balona, L.A.: MNRAS **252**, 93 (1991)
3. Balona, L.A., Cuypers, J., Marang, F.: A&AS **92**, 533 (1992)
4. Balona, L.A.: MNRAS **277**, 1547 (1995)
5. Cranmer, S.R.: ApJ **634**, 585 (2005)
6. Cuypers, J., Balona, L.A., Marang, F.: A&AS **81**, 151(1989)
7. Huat, A.-L., Hubert, A.-M., Baudin, F., et al.: A&A **506**, 95 (2009)
8. Hubrig, S., Yudin, R.V., Pogodin, M., et al.: In: IV Reunion Anual Sociedad Chilena de Astronomia (SOCHIAS), 21 (2007)
9. Meilland, A., Millour, F., Stee, P., et al.: A&A **464**, 73 (2007)
10. Rivinius, T., Baade, D., Štefl, S.: A&A **411**, 229 (2003)
11. Silvester, J., Neiner, C., Henrichs, H.F., et al.: IAU Symp. **259**, 391 (2009)
12. Smith, M.A., Murakami, T., Ezuka, H., et al.: ApJ **481**, 479 (1997)
13. Smith, M.A., Balona, L.A.: ApJ **640**, 491 (2006)
14. Townsend, R.H.D., Owocki, S.P., Howarth, I.D.: MNRAS **350**, 189 (2004)
15. Yudin, R., Hubrig, S., Pogodin, M., et al.: Stars: Be, Polarization, Magnetic Fields, IAU Symposium, vol. 259, p. 397 (2009)

Be Stars: Rapidly Rotating Pulsators

Th. Rivinius

Abstract I will show that Be stars are, without exception, a class of rapidly rotating stars, which are in the majority of cases pulsating stars as well, while none of them does possess a large scale (i.e. with significant dipolar contribution) magnetic field.

1 Introduction

A general discussion of classical Be stars and their properties has been presented by [16]. In this work, I would like to look into results obtained since then with respect to the rotational velocity, the pulsation, and the occurrence of magnetic fields.

To define a Be star, a broad working definition is widely employed: "A non-supergiant B star whose spectrum has, or had at some time, one or more Balmer lines in emission" [5]. However, it must be kept in mind that this definition includes a number of object types that have been awarded their own classes, like interacting binaries of Algol type, stars in which a magnetosphere clearly dominates the emission and its morphology, like σ Ori E, and young stars still accreting out of the surrounding medium or with fossil gaseous disks (for instance 51 Oph, formerly often regarded as classical Be star, has been resorted into that group when its unusual infrared excess became apparent). Therefore, although not useful as a morphological classification criterion, one should note that apart form fulfilling the above definition, only such stars are considered classical Be stars in which this emission arises in a circumstellar disk in Keplerian rotation (at least its outer parts, see [14]), formed by gas ejected by the star itself.

The question fueling the discussion is the following: What is the mechanism responsible for ejecting the material from the stellar surface, and adding sufficient

Th. Rivinius (✉)
ESO - European Organisation for Astronomical Research in the Southern Hemisphere,
Casilla 19001, Santiago 19, Chile
e-mail: triviniu@eso.org

Rivinius, Th.: *Be Stars: Rapidly Rotating Pulsators.*
Astrophys Space Sci Proc. **31**, 253–259 (2013)
DOI 10.1007/978-3-642-29630-7_46,

kinetic energy and angular momentum to it to assume Keplerian rotation, or, as it is usually posed in short: What is the "Be mechanism"?

There was wide consensus that the rapid rotation of Be stars alone, being at about 80% of the critical value ($w = 0.8$, defined as $v_{\rm rot}/v_{\rm crit}$, with tendency to increase towards later types), is not sufficient. Therefore, the discussion about 20 years ago focused on the nature of the short term periodic variations observed both in spectra and integrated light [1], seen in a vast majority of early Be stars, but (then) rarely or not at all in mid- to late-type ones. In the next section, I will review that the consensus on rapid, but not sufficient rotation, has been shaken, and to what extent it might have been restored, in the one following that there should be little doubt left that the short term periodic variations are indeed due to pulsation, and finally that the claim of wide-spread magnetism of Be stars, contributing to the Be-mechanism, does not hold.

2 Rotation

The consensus of rapid, but not sufficient rotation, increasing towards the later type Be stars, has been shaken by the interferometric measurement of the geometrical shape of Achernar [7]. The discussion whether or not a residual disk might have influenced the result was settled with the conclusion that Achernar rotates at $w = 0.99$ of its critical value [4]. It has also been suggested that Be star rotational rates might generally be underestimated, as any value above about $w = 0.8$ would not broaden the lines anymore due to gravity darkening, rendering the equatorial contribution to the line shape insignificant [30].

Notwithstanding, this question is usually investigated employing statistical methods on the measured photospheric line widths and shapes, often based on the compilation by [32], like [6], or FASTROT modeling [8], both including several hundred of stars. However, both approaches come with problems. One problem is the quality of the input data. For instance, upon close inspection several Be stars are not single Be stars, but in fact binary or triple systems where the emission comes from one component, but the photospheric profiles from another. A famous case was β Cep, for some time suspected to be an intrinsically slowly rotating (and magnetic) Be star, unless the binary nature was shown [29]. However, less obvious cases exist as well, such as ν Gem [22] or HR 6819 (Hadrava et al., in prep.). Another concern is the influence of the circumstellar material on the line width and shape. In several cases this has led to published values of $v \sin i$ differing by hundred and more km s^{-1} (like ζ Tau, 27 CMa, κ^1 Aps, and μ For). The six examples above were identified in the HEROS database of 70 Be stars [26, 28], meaning a significant fraction of almost 10% of a given sample might bias the result towards a too early-too slow average when relying on traditional $v \sin i$ statistics.

A more promising approach determines the full set of stellar parameters for a rapidly rotating star (including inclination and equatorial velocity separated) not from line width alone, but shape, hoping to eliminate the gravity darkening problem.

They derive $\bar{w} = 0.75$ [8]. Apart from that it is not entirely clear how sensitive the method is to problems like the biases above, a detailed analysis of the result reveals the certain presence of some own bias: In a histogram of inclinations, the bin from 80° to 90° is almost empty, while for a random distribution (which the rest of the histogram follows well) this should be the most populated one. Also, looking at the inclinations for so-called shell stars (seen edge on through the disk, i.e. the inclination must be quite equatorial), these are fairly wide distributed between about i=50° and i=80°, but again depleted for 80–90. On the other hand, assuming $\sin i = 1$ for shell stars $\bar{w} \approx 0.8$ is obtained for those, but independent of spectral subtype [22].

Finally comparing data from the Galaxy, LMC, and SMC a similar value of $\bar{w}$ as above, for Be star rotation computed back to ZAMS, was found for the Galaxy, but higher for the LMC and higher still for the SMC, again without dependency on spectral type [13].

Putting this all together, there seem to be few things that can be claimed without debate. One is that there is no single Be star that has been shown to rotate really slow, i.e. below about $w = 0.7$ (including uncertainties). As a population, Be stars seem to have around $\bar{w} \approx 0.8$, but given potential biases (like above and [30]), the example of Achernar, and the results from SMC vs. LMC vs. Galaxy, it is more safe to claim this not to be a mean but rather a lower threshold, that increases as metalicity decreases. Whether or not a real trend with spectral subtype exists is unclear.

3 Pulsation

When short term periodic variations were discovered in Be stars, both in photometry and spectroscopy and with periods typically between 0.5 and 1.5 days, the discussion concentrated on whether these should be understood as signature of pulsation or some corotating structure [1]. With increasing quality of spectroscopic data, multiperiodicity in distinct mode groups became apparent [20, 31], and it was possible to model the spectroscopic line profile variations as pulsation in great detail [12, 21] and for a majority of investigated objects [17], while the co-rotation hypothesis, apart from speculation on clouds and differential rotation and a set of toy models, has not yet produced a quantitative model to reproduce the multiperiodicity and the line profile variations. With photometric space missions the multi-periodicity, found in spectroscopic data only with major efforts, was observed to be the rule in Be stars (e.g. [23]).

However, this picture is not complete. Again it was first in spectroscopy that secondary periods were seen, with properties clearly marking them as of circumstellar origin [18, 27] and typically about 10% longer than the photospheric main pulsational periods. In one intensively studied case, μ Cen, these periods were found to last only for about a dozen of cycles and re-occur with slightly different values each time [19]. In this case, they were ascribed to freshly ejected material, undergoing circularization, where the precise properties (i.e. period and phase) depend on the particular properties of the given ejection. In space photometry of Be stars, often two types of peaks are observed in the periodogram. One is narrow and easily understood as the long-term coherent periods corresponding to the stellar pulsation. A second type is typically much broader than the frequency resolution, and often has strong, yet variable contribution from harmonics (e.g. Semaan, this volume). The broad nature of these peaks is not consistent with any corotating feature, but obviously these are the photometric equivalent to the secondary periods seen in spectroscopy.

4 Magnetism

Claims of large scale magnetic fields in the few ten to few hundred Gauss regime have been made for a number of Be stars [10, 11, 15]. Upon close inspection, with one exception these are in the three to four σ regime of significance, which makes it mandatory to have them confirmed independently, in particular since most have been claimed with a single instrument, the reliability of which is debated in the low-significance regime [2, 3, 24, 25]. Indeed, so far two of the seven published claims, for ω Ori and μ Cen, were firmly rejected by the MiMeS collaboration, with none yet confirmed [9]. A re-reduction of the archival data comes to an even stronger conclusion: None of these claims hold [3].

In the MiMeS survey so far 38 Be stars have been probed for the presence of a magnetic field. Not in a single of them was one found. Comparing this to the detection statistics of non-emission B stars probed in the same program (about 8%, with similar significance limit distribution), the conclusion from this is that, in any case, Be stars, as a class, are less magnetic than non-emission B stars.

To this come theoretical problems, as in MHD simulations it was found to be impossible to form a Keplerian disk through magnetic leverage and release at a specified distance, for any degree of parameter-tuning, even though the regime where the angular momentum/kinetic energy balance seems to be correct is well within reach of MHD modeling (Owocki and ud-Doula, private communication).

5 Conclusion

The problem of the Be mechanism is certainly not solved by the above results, which are summarized here:

Rotation: At this point, the statistical analysis of rotational velocities of Be stars is not sufficiently robust to base any firm conclusion on it. A new consensus about Be-star rotation seems to be re-emerging, which in short might be understood as "80% of critical was correct", but a more detailed version of it rather reads like "there is a (metalicity dependent?) threshold rotation, about 80% for the MW, of the critical value, below which no Be stars are formed".

Pulsation: The picture drawn by spectroscopy and space photometry is a very consistent one: There are long-term coherent photospheric pulsation periods, in many cases accompanied by circumstellar quasi-periods signaling the actual mass-transfer from the star to the environment.

Magnetism: Not only no magnetic field in any Be star could be confirmed, but the most thorough search so far, by the MiMeS group, came up with a very clear and significant null-result, at variance to any other class of investigated early-type objects.

As a consequence of the above, a Be mechanism relying on (magnetically) forced corotation on a large scale (i.e. other than highly localized) has no observational support, and as well on theoretical grounds seems to be excluded.

What is it, then? Rapid rotation must certainly be present. In some cases it might suffice by itself, yet in many it doesn't. There is no need to require only one additional mechanism, though. The closer a star rotates to the critical threshold, the easier it is to launch material into a Keplerian orbit (tidal forces, weak pulsation, even turbulence). As we retreat from that limit, less and less mechanisms would get the job done (strong pulsation, pulsational beating pushing amplitudes well above the sound speed for a while), until a threshold is reached, below which no mechanism does work any more. Quite possibly, therefore, the question for the "Be mechanism" has to be recast as the question for the "Be mechanisms".

Acknowledgements I would particularly like to thank the MiMeS collaboration for permission to use the results before or in early stages of publication. Discussions with D. Baade, S. Štefl, S. Owocki, R. Townsend, G. Wade, A. ud-Doula, V. Petit, M. Oksala, C. Neiner, and S. Bagnulo have improved this contribution considerably.

References

1. Baade, D., Balona, L.A.: Periodic variability of Be stars: nonradial pulsation or rotational modulation? Pulsation; rotation; and mass loss in early-type stars. IAU Symp. **162**, 311 (1994)
2. Bagnulo, S., Landstreet, J.D., Fossati, L., Kochukhov, O.: Stellar magnetism through the polarized eyes of the FORS1 instrument. Stellar Polarimetry. In: AIP Conference Proceedings, vol. 1429, pp. 57–66 (2012). arXiv:1109.5043
3. Bagnulo, S., Landstreet, J.D., Fossati, L., Kochukhov, O.: Astron. Astrophys. **545**, 30 (2012)
4. Carciofi, A.C., Domiciano de Souza, A., Magalhães, A.M., Bjorkman, J.E., Vakili, F.: On the determination of the rotational oblateness of Achernar. Astrophys. J. **676**, L41–L44 (2008)
5. Collins, G.W.: II: The use of terms and definitions in the study of Be stars. In: IAU Colloquium 92: Physics of Be Stars. Proceedings of the Ninety-second IAU Colloquium, Boulder, CO, Aug 18–22, 1986 (A88-39926 16-90), pp. 3–19. Cambridge University Press, Cambridge and New York (1987)
6. Cranmer, S.R.: A statistical study of threshold rotation rates for the formation of disks around be stars. Astrophys. J. **634**, 585–601 (2005)

7. Domiciano de Souza, A., Kervella, P., Jankov, S., Abe, L., Vakili, F., di Folco, E., Paresce, F.: The spinning-top Be star Achernar from VLTI-VINCI. Astron. Astrophys. **407**, L47–L50 (2003)
8. Frémat, Y., Zorec, J., Hubert, A.-M., Floquet, M.: Effects of gravitational darkening on the determination of fundamental parameters in fast-rotating B-type stars. Astron. Astrophys. **440**, 305–320 (2005)
9. Grunhut, J.H., Wade, G.A., the MiMeS Collaboration: The incidence of magnetic fields in massive stars: an overview of the MiMeS Survey Component. Stellar Polarimetry. In: AIP Conference Proceedings, vol. 1429, pp. 67–74 (2012). ArXiv e-prints, arXiv:1109.2673
10. Hubrig, S., Yudin, R.V., Pogodin, M., Schöller, M., Peters, G.J.: Evidence for weak magnetic fields in early-type emission stars. Astron. Nachr. **328**, 1133–1136 (2007)
11. Hubrig, S., Schöller, M., Savanov, I., Yudin, R.V., Pogodin, M.A., Štefl, S., Rivinius, T., Curé, M.: Magnetic survey of emission line B-type stars with FORS 1 at the VLT. Astron. Nachr. **330**, 708 (2009)
12. Maintz, M., Rivinius, T., Štefl, S., Baade, D., Wolf, B., Townsend, R.H.D.: Stellar and circumstellar activity of the Be star ω CMa, III. Multiline non-radial pulsation modeling. Astron. Astrophys. **411**, 181–191 (2003)
13. Martayan, C., Frémat, Y., Hubert, A.-M., Floquet, M., Zorec, J., Neiner, C.: Effects of metallicity, star-formation conditions, and evolution in B and Be stars, II. Small Magellanic Cloud, field of NGC 330. Astron. Astrophys. **462**, 683–694 (2007)
14. Meilland, A., Millour, F., Kanaan, S., Stee, P., Petrov, R., Hofmann, K.-H., Natta, A., Perraut, K.: First spectro-interferometric survey of Be stars I. Observations and constraints on the disks geometry and kinematics. Astron. Astrophys. **538**, 110 (2012). arXiv:1111.2487
15. Neiner, C., Hubert, A.-M., Frémat, Y., Floquet, M., Jankov, S., Preuss, O., Henrichs, H.F., Zorec, J.: Rotation and magnetic field in the Be star ω Orionis. Astron. Astrophys. **409**, 275–286 (2003)
16. Porter, J.M., Rivinius, T.: Classical Be stars. Publ. Astron. Soc. Pac. **115**, 1153–1170 (2003)
17. Rivinius, T., Baade, D., Štefl, S.: Non-radially pulsating Be stars. Astron. Astrophys. **411**, 229–247 (2003)
18. Štefl, S., Baade, D., Rivinius, T., Stahl, O., Wolf, B., Kaufer, A.: Circumstellar quasi-periods accompanying stellar periods of Be stars. In: Bradley, P.A., Guzik, J.A. (eds.) A Half Century of Stellar Pulsation Interpretation, vol. 135. PASPC, p. 348 (1998)
19. Rivinius, T., Baade, D., Štefl, S., Stahl, O., Wolf, B., Kaufer, A.: Stellar and circumstellar activity of the Be star μ Centauri, I. Line emission outbursts. Astron. Astrophys. **333**, 125–140 (1998)
20. Rivinius, T., Baade, D., Štefl, S., Stahl, O., Wolf, B., Kaufer, A.: Stellar and circumstellar activity of the Be star MU Centauri, II. Multiperiodic low-order line-profile variability. Astron. Astrophys. **336**, 177–190 (1998)
21. Rivinius, T., Baade, D., Štefl, S., Townsend, R.H.D., Stahl, O., Wolf, B., Kaufer, A.: Stellar and circumstellar activity of the Be star μ Centauri, III. Multiline nonradial pulsation modeling. Astron. Astrophys. **369**, 1058–1077 (2001)
22. Rivinius, T., Štefl, S., Baade, D.: Bright Be-shell stars. Astron. Astrophys. **459**, 137–145 (2006)
23. Semaan, T., Gutiérrez-Soto, T., Frémat, Y., Hubert, A.M., Martayan, C., Zorec, J.: A pulsational study of a sample of CoRoT Faint Be Stars. In: Suárez, J.C., Garrido, R., Balona, L.A., Christensen-Dalsgaard, J. (eds.) Stellar Pulsations. Impact of New Instrumentation and New Insights. Springer, Heidelberg (2012)
24. Shultz, M., Wade, G., Bagnulo, S., Landstreet, J., Grunhut, J., Hanes, D., the MiMeS Collaboration: Magnetic fields of slowly pulsating B stars and β Cep variables: comparing results from FORS1/2 and ESPaDOnS. Stellar Polarimetry. In: AIP Conference Proceedings, vol. 1429, pp. 110–113 (2012). arXiv:1109.5436
25. Silvester, J., Neiner, C., Henrichs, H.F., et al.: On the incidence of magnetic fields in slowly pulsating B, β Cephei and B-type emission-line stars. Mon. Not. R. Astron. Soc. **398**, 1505–1511 (2009)
26. Štefl, S., Rivinius, T.: HEROS Be star campaigns. In: Smith, M.A., Henrichs, H.F. (eds.) IAU Colloquium 175: The Be Phenomenon in Early-Type Stars, vol. 214, p. 356 (2000)

27. Štefl, S., Baade, D., Rivinius, T., Stahl, O., Budovičová, A., Kaufer, A., Maintz, M.: Stellar and circumstellar activity of the Be star ω CMa, II. Periodic line-profile variability. Astron. Astrophys. **411**, 167–180 (2003)
28. Štefl, S., Rivinius, T.: Be star research by the HEROS group: results from the past decade. Publ. Astron. Inst. Czech. Acad. Sci. **93**, 1–8 (2005)
29. Schnerr, R.S., Henrichs, H.F., Oudmaijer, R.D., Telting, J.H.: On the Hα; emission from the β Cephei system. Astron. Astrophys. **459**, L21–L24 (2006)
30. Townsend, R.H.D., Owocki, S.P., Howarth, I.D.: Be-star rotation: How close to critical? Mon. Not. R. Astron. Soc. **350**, 189–195 (2004)
31. Tubbesing, S., Rivinius, T., Wolf, B., Kaufer, A.: Multiperiodic variability and outbursts of 28 Cygni. In: Smith, M.A., Henrichs, H.F. (eds.) IAU Colloquium 175: The Be Phenomenon in Early-Type Stars, vol. 214, p. 232 (2000)
32. Yudin, R.V.: Statistical analysis of intrinsic polarization, IR excess and projected rotational velocity distributions of classical Be stars. Astron. Astrophys. **368**, 912–931 (2001)

A Pulsational Study of a Sample of *CoRoT* Faint Be Stars

T. Semaan, J. Gutiérrez-Soto, Y. Frémat, A.M. Hubert, C. Martayan, and J. Zorec

Abstract We present an intensive study of a sample of 12 Be stars in the first exoplanetary fields of the *CoRoT* mission. The aim is to understand the short-term variability in Be stars. We use both spectroscopic and photometric data: VLT-GIRAFFE spectra to derive the fundamental parameters and *CoRoT* light curves to search for low-amplitude frequencies. This allows us to locate the stars in the HR diagram and compare their positions in the instability strip with those of other pulsating B stars. From VLT-GIRAFFE observations we determined the stellar parameters by fitting the observed spectra with non-LTE stellar atmosphere models. The observed spectra were corrected for veiling effects due to circumstellar material and for rotational effects. We estimated the rotational frequency using the corrected fundamental parameters. For the *CoRoT* light curves, we determined all significant frequencies and compared the main frequencies with the rotational frequency. We also investigated the variation of the frequencies and amplitudes with time and, in particular, during outbursts. This is the first time that a statistical study with space

Semaan, T. (✉) · Hubert, A.M.
GEPI, Observatoire de Paris, France
e-mail: thierry.semaan@obspm.fr; anne-marie.hubert@obspm.fr

Gutiérrez-Soto, J.
Valencian International University & Instituto de Astrofisica de Andalucia (CSIC), Spain
e-mail: jgs@iaa.es

Frémat, Y.
Royal Observatory of Belgium, Belgium
e-mail: yves.fremat@oma.be

Martayan, C.
ESO, Chile
e-mail: cmartaya@eso.org

Zorec, J.
UPMC, CNRS, UMR 7095, Institut d'Astrophysique de Paris, France
e-mail: zorec@iap.fr

Semaan, T. et al.: *A Pulsational Study of a Sample of* CoRoT *Faint Be Stars.* 261
Astrophys Space Sci Proc. **31**, 261–264 (2013)
DOI 10.1007/978-3-642-29630-7_47,

data has been performed for Be stars. Our study suggests that rotational modulation is not the cause of short-term photometric variability in Be stars.

1 Introduction

The origin of short-term photometric variations in Be stars is still not well understood [8]. Two main hypotheses are currently evoked to explain the variations: (a) rotational modulation by transits of spots on the stellar surface or (b) non-radial pulsation. In order to clarify these questions, we have begun a study of light curves of a significant number of Be stars in the first exo-planetary fields of *CoRoT* (IR1, LRA1, LRC1). One of the objectives is to identify what are the common properties of the short-term photometric variations of Be stars, and how they change according to position in the HR diagram.

2 Determination of Fundamental Parameters and Estimation of the Rotational Frequency

Thanks to several photometric and spectroscopic surveys, we have identified more than 30 Be stars in the first exo-planetary fields of *CoRoT* (IR1, LRA1, LRC1). However, during 2009 and 2010 we were able to obtain spectroscopic FLAMES/GIRAFFE data (Program ID: 082.D-0839(A)) for only 15 of these stars. We used the LR2 setup (resolution $R = 6,400$) in the 4,000–4,500 Å wavelength range. These spectra enable us to estimate T_{eff}, $\log g$ and $V \sin i$. To this end, we used the `GIRFIT` code [6] which is a robust least-square minimization program to fit observed spectra with non-LTE plane-parallel model atmospheres. Since Be stars are fast rotators, the fundamental parameters thus obtained need to be corrected for rotational effects to obtain the true stellar mass and evolutionary status. These corrections take into account effects induced by rotation on the stellar structure and its geometry, as well as the surface gravity darkening. The corrections were obtained with the `FASTROT` [5] code. They were estimated for fixed values of the angular velocity to critical velocity ratio, $\Omega/\Omega_{\mathrm{c}}$, ranging from 0.8 to 1.0.

Once T_{eff}, $\log g$ and $V \sin i$, corrected for the appropriate value of $\Omega/\Omega_{\mathrm{c}}$, have been derived, the bolometric luminosity and effective temperature appropriate to a non-rotating counterpart can be obtained. These parameters enable us to infer a first estimate of the true stellar mass. On this basis, we then re-calculate the surface-averaged bolometric luminosity and effective temperature for the given $\Omega/\Omega_{\mathrm{c}}$ [10]. These quantities, together with evolutionary tracks for rotating stars [4, 7], enable the true stellar mass and age to be estimated. They also enable estimates of the equatorial radius of the rotationally-deformed star. This value, together with $V \sin i$ corrected for gravity darkening (which leads to an underestimate of the rotational velocity if not used [9]) and for the inclination angle, gives an estimate of the rotational frequency of the star.

3 Determination of Pulsation Frequencies

We determined the pulsation frequencies for each star from the *CoRoT* light curves [1]. The time duration of the light curves depends on the observation run which is about 130 days for the LRA1 and LRC1 fields. This leads to a resolution frequency of 0.008 d^{-1}. For the IR1 field the duration is about 60 days, or a resolution frequency of 0.017 d^{-1}. The frequencies were derived iteratively using the Fourier transform of the light curve and fitting a sinusoid to the frequency of highest-amplitude in the original light curve [3]. This frequency is pre-whitened and further frequencies are derived from the residuals. The procedure is repeated until the highest-amplitude frequency has an amplitude smaller than 3.5 times the noise in the frequency spectrum. We also performed a time-frequency analysis to determine how the amplitude changes with time.

4 Results

- In most of the Be stars that we studied, the frequencies are distributed in two or three distinct groups.
- Several Be stars have very close frequencies of high amplitude which lead to beats clearly seen in the light curve.
- Be stars that are slightly evolved from the main sequence show a poor frequency spectrum: less than 20 frequencies with very low amplitudes (350 ppm) are detected in these stars.
- Be stars in the main sequence, on the contrary, show many frequencies, from 60 to 100, although some of them are combinations of the higher amplitude frequencies.
- Several Be stars show outbursts in their light curves. The amplitude variations of the main frequencies during the outburst have similar behaviour.
- In order to determine if rotational modulation is a possible explanation of the short-term photometric variations of Be stars, we compared the five highest-amplitude frequencies with the rotational frequency calculated for fixed values of Ω/Ω_c ranging from 0.8 to 1.0. In most cases the observed frequencies lie outside the estimated rotational frequency interval. However, in two Be stars that show clear outbursts, a frequency that matches the rotational frequency does appear during the outburst.
- Finally, we determined the frequency spacings (as in [2]) for each *CoRoT* Be star. We see that for most of the stars, the frequencies are distributed in groups with a frequency that, according to [2], should be the rotational frequency. Figure 1 shows that, in most of the cases, there is no match. In all Be stars studied here, the estimated rotational frequencies derived from fundamental parameters are lower than those obtained from the frequency spacings.

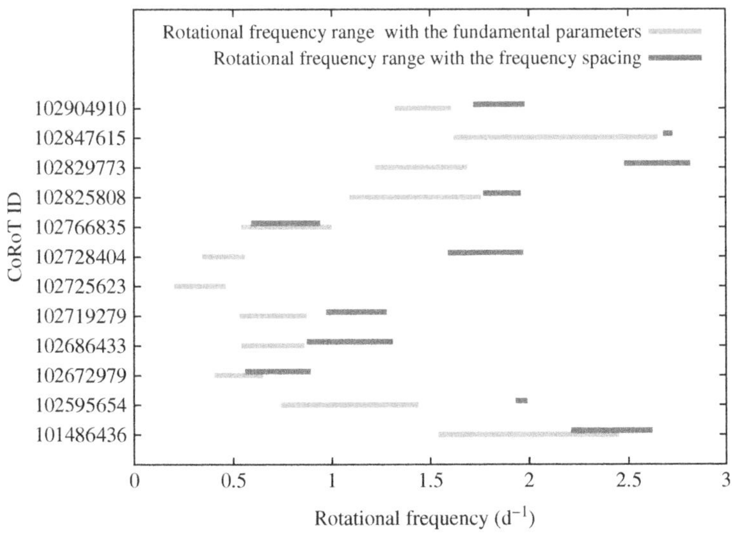

Fig. 1 Comparison between the observed frequency spacings and the rotational frequency intervals calculated with spectroscopy

Acknowledgements This research is based on *CoRoT* data. The *CoRoT* (Convection, Rotation and planetary Transits) space mission, launched on December 27th 2006, has been developed and is operated by CNES, with the contribution of Austria, Belgium, Brazil, ESA, Germany and Spain.

References

1. Auvergne, M., Bodin, P., Boisnard, L., et al.: A&A **506**, 411 (2009)
2. Balona, L.A., Pigulski, A., Cat, P.D., et al.: MNRAS **413**, 2403 (2011)
3. Diago, P.D., Gutiérrez-Soto, J., Auvergne, M., et al.: A&A **506**, 125 (2009)
4. Ekström, S., Meynet, G., Maeder, A., Barblan, F.: A&A **478**, 467 (2008)
5. Frémat, Y., Zorec, J., Hubert, A.-M., Floquet, M.: A&A **440**, 305 (2005)
6. Frémat, Y., Neiner, C., Hubert, A.-M., et al.: A&A **451**, 1053 (2006)
7. Meynet, G., Maeder, A.: A&A **361**, 101 (2000)
8. Porter, J.M., Rivinius, T.: PASP **115**, 1153 (2003)
9. Stoeckley, T.R.: MNRAS **140**, 149 (1968)
10. Zorec, J., Frémat, Y., Cidale, L.: A&A **441**, 235 (2005)

Mode Identification for Subdwarf B Stars Using Period Spacings in *Kepler* Data

M.D. Reed

Abstract During the *Kepler* spacecraft's first year of operation, its short cadence observations were obtained in a survey mode where targets received one month of nearly continuous observations. Forty eight subdwarf B stars were observed of which 14 were found to be pulsators, 13 having predominantly long (g mode) periods. We applied asymptotic relations to the g-mode pulsations and discovered evenly spaced in period sequences for $\ell = 1$ and 2 modes in all of the stars. We were able to associate 222 of 317 measured periods with $\ell = 1$ and 2 modes which should provide tight constraints on pulsation models. We use the period spacings to compare with theoretical masses of these stars.

1 Introduction

Subdwarf B (sdB) variables were first discovered in 1996 and now consist of two well-established classes (for a review see [12]): short period (90–600 s), and long period (45 min to 2 h). The short-period pulsators are likely to be pressure (p-mode) pulsators which are designated V361 Hya stars [8] and the long-period pulsators are likely to be gravity (g-mode) pulsators which are designated V1093 Her stars [3].

Several papers in these proceedings (see De Ridder et al., Hekker et al., Garcia et al., Antoci et al., among others) discuss modeling the internal mechanisms of stars. However, the best results are usually achieved for stars where pulsation periods (or frequencies) are observationally associated with pulsation modes (see the papers of Pamyatnykh et al., Guzik et al., Suárez et al., these proceedings). For g-modes in compact pulsators, equally spaced periods indicate consecutive overtones for $n \gg \ell$. These are the well-known asymptotic relations of Ledoux [9].

M.D. Reed (✉)
Missouri State University, 901 S. National, Springfield, MO 65897, USA
e-mail: mikereed@missouristate.edu

Reed, M.D.: *Mode Identification for Subdwarf B Stars Using Period Spacings in* Kepler *Data*. Astrophys Space Sci Proc. **31**, 265–268 (2013)
DOI 10.1007/978-3-642-29630-7_48,

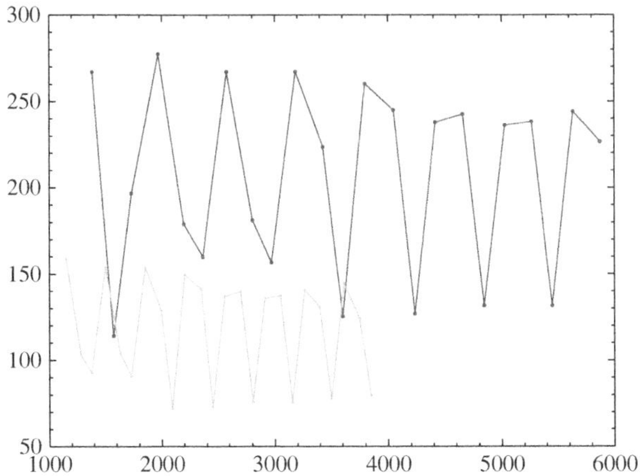

Fig. 1 Sample evolutionary model of the prototype V1093 Her star, PG 1716+426 [13] showing significant mode trapping. Top set of connected points are $\ell = 1$ periods and bottom set are for $\ell = 2$

Working group 11 of the Kepler Asteroseismic Science Consortium published results indicating that *all* of the *Kepler*-observed V1093 Her stars have periods matching the asymptotic relations for $\ell = 1$ and most have periods matching those for $\ell = 2$ also [14]. Here we discuss the significance of evenly spaced periods on mode identifications, stellar models, and masses. For a review of how the period spacings were detected, see Reed et al. [14, 15].

2 Model Comparisons

From past experience with g-mode periods in pulsating sdB stars, we knew to expect poor model matching because of strong mode trapping. Figure 1 shows an evolutionary pulsation model we produced to evaluate the first V1093 Her star discovered, PG 1716+426 [13]. Strong mode trapping is evident and similar results were found by [1] using static atmospheric models, though they only examined $\ell = 3$ modes for sdB stars hotter than the V1093 Her-class.

Two of the *Kepler* stars were fitted with pulsation models , so we can directly compare results. For KIC 2697388 [2], 52% of the 36 mode identifications disagree and for KIC 5807616 [16], the disagreement is 45% for the 22 periods. These disagreements indicate that something is missing from the models, something which would smooth out the hard boundary and lessen mode trapping. One possibility is diffusion. Work by [6, 7, 10] show that diffusion smears out the boundary and greatly decreases mode trapping. Future work in this direction, using evolutionary diffusion models will likely be fruitful since we cannot currently trust mode identifications using period matching to models only.

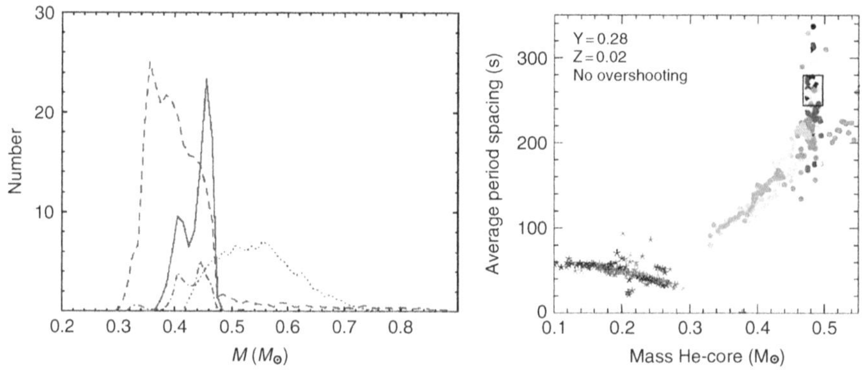

Fig. 2 *Left panel*: Resultant sdB masses via binary evolution. The *solid line* indicates the mass distribution from the first common-envelope ejection channel, the *dot-dashed line* that from the second common-envelope ejection channel, the *dashed line* from the first Roche lobe overflow channel, and the *dotted line* from the helium-helium white dwarf merger channel [4, 5]. *Right panel*: Model core helium masses correlated with period spacings for red giant stars [11]. The *box* indicates possible masses for the range of observed period spacings in *Kepler* pulsating sdB stars

3 Discussion

An interesting prospect is the connection to other stars along the horizontal branch. Red clump stars are metal-rich core helium burning stars with thick envelopes whereas sdB stars are hot core helium burning stars with exceedingly thin, and therefore inert envelopes. These stars are at opposite ends of the horizontal branch, yet they have the same period spacings, indicating that we are probing the same core in both cases.

Another intriguing feature of sdB stars are their incredibly thin envelopes. Models estimate the envelopes at $<10^{-2} M_{\odot}$ which is too thin for shell hydrogen fusion. The question has been, where did the mass go between the tip of the red giant branch, where it was necessary for commencing core helium fusion, and the extended horizontal branch? Most of the proposed solutions involve binary evolution where the companion helps to strip off nearly all of the envelope [4,5]. The left panel of Fig. 2 is from [4] and indicates the mass distribution created by different binary mass ejections. The figure illustrates that different methods for removing the envelope create different sdB mass distributions. So if we can constrain the mass of the sdB, we constrain the mechanism for making it. The right panel of Fig. 2 shows a distribution of core masses correlated with their period spacings produced by [11]. The black rectangle indicates the range of period spacings observed in *Kepler* sdB stars. This correlation constrains the core mass to $0.45 < M_{\odot} < 0.49$ and since sdB stars are essentially all stars, we can directly apply that to the mass distribution of [4]. According to these results, the pulsating *Kepler* sdB stars were not produced either via the first Rocha lobe overflow channel or the helium-helium white dwarf merger channel. The only channels remaining are the common-envelope ones.

References

1. Charpinet, S., et al.: ApJS **139**, 487 (2002)
2. Charpinet, S., et al.: A&A **530**, 3 (2011)
3. Green, E.M., et al.: ApJL **583**, 31 (2003)
4. Han, Z., Podsiadlowski, P., Maxted, P.F.L., et al.: MNRAS **336**, 449 (2002)
5. Han, Z., Podsiadlowski, Ph., Maxted, P.F.L., et al.: MNRAS **341**, 669 (2003)
6. Hu, H., Nelemans, G., Aerts C., et al.: A&A **508**, 869 (2009)
7. Hu, H., Glebbeek, K, Thou, A.A., et al.: A&A **511**, 87 (2010)
8. Kilkenny, D., Koen, C., O'Donoghue, D., et al.: MNRAS **285**, 640 (1997)
9. Ledoux, P.: ApJ **114**, 373 (1951)
10. Miller Bertolami, M.M., Córsico, A.H., Althaus, L.G.: ASP Conf. Ser. **452**, 175 (2012)
11. J. Montalban, A. Miglio, A. Noels, R. Scuflaire, et al.: Proc. of the workshop "Red Giants as Probes of the Structure and Evolution of the Milky Way. In: Miglio, A., Montalban, J., Noels, A. (eds.) Astrophysics and Space Science Proceedings, ISBN 978-3-642-18417-8 (2010)
12. Østensen, R.H.: Astron. Nachr. **331**, 1029 (2010a)
13. Reed, M.D., et al.: ApJ **607**, 445 (2010)
14. Reed, M.D., et al.: MNRAS **414**, 2885 (2011)
15. Reed, M.D., et al.: ASP Conf. Ser. **452**, 193 (2012)
16. Van Grootel, V., et al.: ApJL **718**, 97 (2010)

Pulsating Pre-Main Sequence Stars in NGC 2264 Discovered by the *MOST* and *CoRoT* Satellites

K. Zwintz, L. Fossati, T. Ryabchikova, and T.G. Barnes

Abstract The satellites *MOST* and *CoRoT* have observed the young open cluster NGC 2264 in 2006 and 2008, respectively. The resulting high-precision time-series photometry has been used to investigate the pulsational properties of pre-main sequence (PMS) cluster members. We present a highlight of this study.

1 The Young Cluster NGC 2264

Numerous studies in the literature are dedicated to the young open cluster NGC 2264 and its members. Its age is reported to lie between three and ten million years [11, 13]. With such a young age, the cluster's main sequence consists only of very massive O and B type stars while stars of later spectral types are still contracting towards the main sequence, have not started hydrogen core burning and hence are in their pre-main sequence (PMS) phase. NGC 2264 also hosts the very first discovered

K. Zwintz (✉)
Institute of Astronomy, University of Vienna, Türkenschanzstrasse 17, A-1180 Vienna, Austria
e-mail: konstanze.zwintz@univie.ac.at

L. Fossati
Department of Physics and Astronomy, The Open University, Walton Hall, Milton Keynes MK7 6AA, UK
e-mail: l.fossati@open.ac.uk

T. Ryabchikova
Institute of Astronomy, Russian Academy of Sciences, Pyatnitskaya 48, 119017 Moscow, Russia
e-mail: ryabchik@inasan.ru

T.G. Barnes
The University of Texas at Austin, McDonald Observatory, 82 Mt. Locke Rd., McDonald Observatory, Texas 79734, USA
e-mail: tgb@astro.as.utexas.edu

Zwintz, K. et al.: *Pulsating Pre-Main Sequence Stars in NGC 2264 Discovered by the* MOST *and* CoRoT *Satellites*. Astrophys Space Sci Proc. **31**, 269–272 (2013)
DOI 10.1007/978-3-642-29630-7_49,

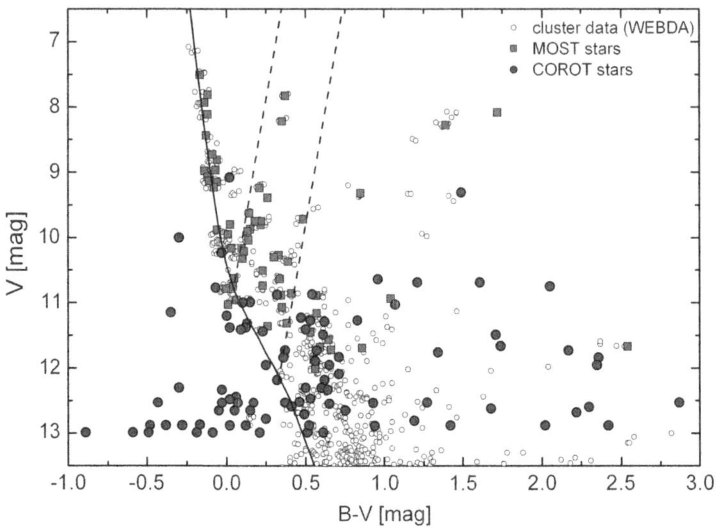

Fig. 1 Colour-magnitude diagram for NGC 2264 with the approximate location of the δ Scuti instability strip (*dashed lines*). Cluster stars are taken from the WEBDA database (*open circles*). Stars observed by *MOST* in 2006 are shown by (*red*) *squares*; stars observed by *CoRoT* in 2008 by (*blue*) *filled circles*

PMS δ Scuti type stars, V588 Mon and V589 Mon [3]. Several observations of NGC 2264 have been conducted over the last few years with the aim of searching for pulsation among cluster members, to determine the respective stars' pulsational characteristics [17] and to conduct asteroseismic analyses [18].

2 NGC 2264 Space Observations

In 2006, the Canadian *MOST* satellite [15] observed 68 stars in the field of NGC 2264 for about 23 days [17]. Using these observations, four new PMS δ Scuti type stars were discovered [17]. In 2008, the European satellite *CoRoT* [1] obtained high-precision photometric time series for 636 stars in the region of NGC 2264. Thirteen new (PMS) pulsators and ten candidates for (PMS) pulsation, which are δ Scuti, γ Doradus and δ Scuti—γ Doradus hybrid stars, were discovered. Additionally, the two known PMS δ Scuti stars, V588 Mon and V589 Mon, were observed with both satellites.

The optimum magnitude range for observations of stars in the open field of the *MOST* satellite is from 7 to 12 mag in V, while the corresponding magnitude range in the exo-field of the *CoRoT* satellite is from 10 to 16 mag in R (see Fig. 1). The combination of data from both satellites provides an unique opportunity to conduct a homogeneous asteroseismic study of stars in the cluster. Here we describe a highlight of the latest results of this study.

2.1 *CoID0500007018 and CoID0223979759*

CoID0500007018 (NGC 2264 VAS 20) and CoID0223979759 (NGC 2264 VAS 87) were observed with the *CoRoT* satellite in 2008. For CoID0500007018 a spectral type of F5 is known in the literature [16]. Using ROSAT observations, an X-ray flux was detected [4] indicating the star's young age. The star is also a cluster member according to its radial velocity [5]; its membership probability of NGC 2264 is reported to be 94% [14]. For CoID0223979759 a spectral type of F9 is reported in the literature [16]. Its membership probability is 96% [14] and it is an X-ray source which was detected by the CHANDRA satellite [9].

A frequency analysis using the software package Period04 [8] yielded 11 and 7 intrinsic frequencies for CoID0500007018 and CoID0223979759, respectively. The clear presence of several independent frequencies rules out rotation as the *only* cause of variability, although rotation might contribute to the observed variations. We therefore conclude that the stars are pulsational variables.

Using high-resolution spectra obtained with the Robert G. Tull Coudé Spectrograph (TS) on the 2.7-m telescope of Mc Donald Observatory, we derived the fundamental parameters and chemical abundances for the two stars (for more details see [19]). Both stars have effective temperatures of $6,300 \pm 200$ K. We measured $v \sin i = 42 \pm 2\,\mathrm{km\,s^{-1}}$ and $\log g = 3.9 \pm 0.2$ for CoID050007018, and $v \sin i = 18 \pm 1\,\mathrm{km\,s^{-1}}$ and $\log g = 3.8 \pm 0.2$ for CoID0223979759. The chemical composition of both stars is comparable to the sun except for a clear Li overabundance. For Li we derived abundances of log n(Li)=3.34±0.16 for CoID0500007018 and log n(Li)=3.54±0.10 for CoID0223979759. These values are consistent with a study conducted for other stars of similar effective temperatures in the cluster [12].

Combining the high-precision photometric time series obtained with the *CoRoT* satellite with the high-resolution spectroscopy from the ground, we therefore conclude that CoID0500007018 and CoID0223979759 are the first two members of the group of PMS γ Doradus stars. The existence of this group of objects was predicted by theory [2], but until now no confirmed members of this group have been reported. Future work will focus on the asteroseismic interpretation of these two stars and their comparison to their more evolved (post-)main sequence counterparts.

3 Summary of Pulsating Pre-main Sequence Stars

Ten years ago only eight PMS δ Scuti type pulsators were known. They were thought to be purely radial pulsators.

As of September 2011, more than 60 pulsating PMS stars are known. They are either members of young open clusters or belong to the group of Herbig Ae/Be stars. With the present work the first two members of a new class of pulsators among PMS objects were discovered: the g-mode pulsators PMS γ Doradus type stars.

PMS stars are found to be radial and non-radial pulsators for which the respective pulsation theory was developed during the last few years (e.g., [6, 7, 10]). The first attempt to use asteroseismology as an independent method to put constraints on the evolutionary stage of a star has already been a success [6].

From December 2011 to January 2012 a multi-satellite campaign involving the *MOST*, *CoRoT*, *Spitzer* and *CHANDRA* satellites will be conducted. All four space telescopes will (re-)observe the young open cluster NGC 2264 simultaneously. Additionally, a large ground-based campaign of different supportive observations is planned. With these new observations new insights in the study of young stars, their evolution and their seismic properties are expected.

Acknowledgements KZ is recipient of an APART fellowship of the Austrian Academy of Sciences at the Institute of Astronomy of the University Vienna. Astronomy research at the Open University is supported by an STFC rolling grant (L.F.). We also acknowledge the use of cluster facilities at the Institute of Astronomy of the University of Vienna. Data were obtained with the 2.7-m telescope at Mc Donald Observatory, Texas, US.

References

1. Baglin, A.: The *CoRoT* Mission, Pre-launch Status, Stellar Seismology and Planet Finding. ESA SP-1306, ESA Publication Division, Noordwijk, The Netherlands (2006)
2. Bouabid, M.-P., et al.: A&A **531**, 145 (2001)
3. Breger, M.: ApJ **171**, 539 (1972)
4. Flaccomio, E., et al.: A&A **355**, 651(200)
5. Fűrész, G., et al.: ApJ **648**, 1090 (2006)
6. Guenther, D.B., et al.: ApJ **671**, 581(2007)
7. Guenther, D.B., et al.: ApJ **704**, 1710 (2009)
8. Lenz, P., Breger, M.: CoAst **146**, 53 (2005)
9. Ramírez, S.V., et al.: AJ **127**, 2659 (2004)
10. Ruoppo, A., et al.: A&A **466**, 261 (2007)
11. Sagar, R., et al.: MNRAS **220**, 383 (1986)
12. Sestito, P., Randich, S.: A&A **442**, 615 (2005)
13. Sung, H., Bessel, M.S., Chun, M.-Y.: AJ **128**, 1684 (2004)
14. Vasilevskis, S., Sanders, W.L., Balz, A.G.A.: AJ **70**, 797 (1965)
15. Walker, G., et al.: PASP **115**, 1023 (2003)
16. Young, A.: PASP **90**, 144 (1978)
17. Zwintz, K., et al.: A&A **502**, 239 (2009)
18. Zwintz, K., et al.: ApJ **729**, 20 (2011)
19. Zwintz, K., et al.: ApJL (submitted)

Constraints on Pasta Structure of Neutron Stars

Hajime Sotani

Abstract We show that shear modes in a neutron star crust are quite sensitive to the existence of non-uniform nuclear structures, the so-called "pasta." Due to the existence of the pasta phase, the frequencies of shear modes are reduced and the dependence on fundamental frequency is different from that on overtones. Since the torsional shear frequencies depend strongly on the structure of the pasta phase, one can probe the pasta structure in the crust from observations of stellar oscillations. This is difficult to do with other types of observations. In addition, by considering the effect of the pasta phase, we show the that it is possible to explain the observed frequencies in SGR 1806-20 using only crust torsional shear modes.

1 Introduction

The soft gamma repeaters (SGRs) are considered as one of the most promising candidates for magnetars, which are neutron stars with strong magnetic fields. Sporadic X- and gamma-ray bursts are radiated from SGRs, while SGRs rarely emit the much stronger gamma-rays called "giant flares." Up to now, at least three giant flares have been detected. Through timing analyses of the decaying tail, quasi-periodic oscillations (QPOs) have been discovered. These are in the range from tens of Hz up to a few kHz [1]. In order to understand these QPO frequencies, many theoretical calculations have been made on oscillations in magnetars [2–5]. Unfortunately, none of these attempts have yet been successful. In these calculations it is assumed that the nuclei form a body-center cubic (bcc) lattice in the crust of the neutron star. However, the existence of a non-uniform nuclear structure at the bottom

H. Sotani (✉)
Division of Theoretical Astronomy, National Astronomical Observatory of Japan, 2-21-1 Osawa, Mitaka, Tokyo 181-8588, Japan
e-mail: sotani@astro.auth.gr

Sotani, H.: *Constraints on Pasta Structure of Neutron Stars*.
Astrophys Space Sci Proc. **31**, 273–276 (2013)
DOI 10.1007/978-3-642-29630-7_50, © Springer-Verlag Berlin Heidelberg 2013

of the solid crust in a neutron star, the so-called "pasta," is well-known. Unlike the case of a bcc lattice, the shear modulus may decrease with increasing density in the pasta phase [6]. If so, the oscillation frequencies of magnetars could be lower than in the case of a bcc lattice structure, because the shear speed should be lower if the shear modulus is lower. In this article, we adopt the simple shear modulus in the pasta phase as a first step and compare the frequencies of shear torsional oscillations with the observed QPOs. For more details of this study see [7].

2 Numerical Results

Although nuclei in the crust form a bcc lattice due to Coulomb interactions, recent studies show that the nuclear structure in the bottom of the crust could be nonuniform. In other words, as the density increases, the shape of nuclear matter in this region changes from a sphere (bcc lattice) to a cylinder, slab, cylindrical hole, and uniform matter (inner fluid core). The density where the cylinder structure appears, ρ_p, depends on the nuclear symmetry energy which is expressed with density symmetry coefficient L and is suggested to be of order $\rho_p \sim 10^{13}\,\mathrm{g\,cm^{-3}}$ from calculations of the ground state of matter in the crust. With the assumption that nuclear matter forms a bcc lattice, the shear modulus is suggested to be

$$\mu = 0.1194 \times n_i (Ze)^2/a, \tag{1}$$

where n_i is the ion number density, $a = (3/4\pi n_i)^{1/3}$ is the average ion spacing, and $+Ze$ is the ion charge. On the other hand, it has been pointed out that the elastic properties in the pasta phase could be that of a liquid crystal rather than a crystalline solid [6]. Thus the shear modulus should decrease with increasing density in the pasta phase. This picture is thought to be natural, because the structure of nuclear matter changes gradually, as mentioned above, and the shear modulus will finally become zero in the fluid core. In order to obtain such a relation for the shear modulus, we adopt Eq. (1) in the crust except for the pasta phase. In the pasta phase it is assumed that the shear modulus can be expressed as a cubic function of the density. This satisfies the condition that μ should connect smoothly to Eq. (1) at $\rho = \rho_p$ and should tend to zero smoothly at the core boundary, i.e., $\mu = c_1(\rho - \rho_c)^2(\rho - c_2)$, where c_1 and c_2 are constants determined by the boundary conditions (see Fig. 1). Of course, this simple function of μ in the pasta phase might be a kind of toy model, although it does express roughly the correct behaviour. In reality, this should depend on the microscopic structure which includes the composition and/or the nuclear symmetry energy. Nevertheless, this relationship is probably adequate as a first step in examining the dependence of shear oscillations in the presence of the pasta phase.

We examine the frequencies of torsional shear modes as a function of ρ_p in the range $\rho_p = 10^{13}$–$10^{14}\,\mathrm{g\,cm^{-3}}$, and compare the calculated frequencies of shear modes with the observed QPO frequencies in a giant flare. In particular, we focus

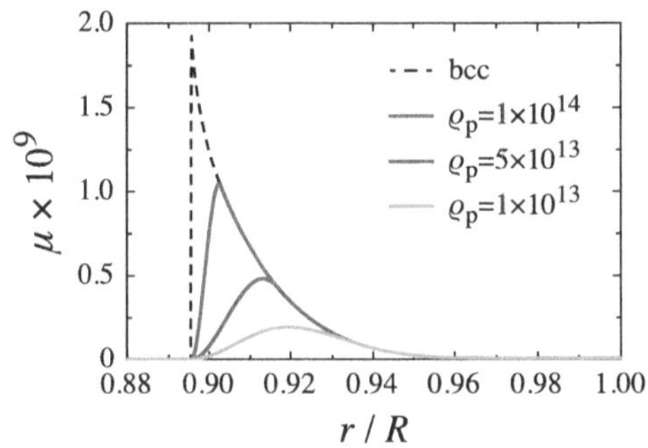

Fig. 1 Adopted shear modulus μ as a function of relative radius r/R

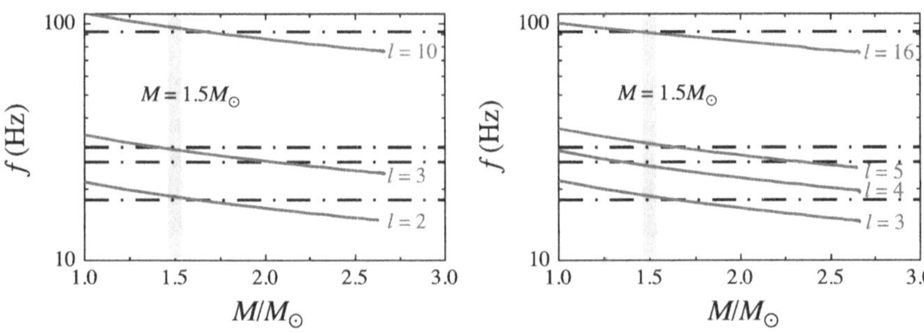

Fig. 2 Comparison of calculated frequencies with the observed QPO frequencies in SGR 1806-20 with $\rho_p = 10^{14}\,\mathrm{g\,cm^{-3}}$ (*left*) and with $\rho_p = 2 \times 10^{13}\,\mathrm{g\,cm^{-3}}$ (*right*)

in this article on the QPO frequencies in SGR 1806-20 which are under 100 Hz, i.e., 18, 26, 30, and 92.5 Hz. This is because these are the most commonly observed frequencies among giant flares detected in the past [1]. Previous studies have shown the difficulty of explaining all the observed frequencies using only crust shear modes [2]. The fundamental $\ell = 2$ mode, which has the lowest possible frequency, is thus considered to correspond to the observed frequency of 18 Hz while 26 and 30 Hz corresponds to the fundamental $\ell = 3$ mode. However, the spacing of shear frequencies for different ℓ is larger than the spacing between the observed frequencies of 26 and 30 Hz. In practice, according to such a traditional identification, we can explain the observed frequencies as crust shear modes in the pasta phase as shown in the left panel of Fig. 2. Thus the 18, 30 and 92.5 Hz modes can be identified as the $\ell = 2$, 3, and 10 fundamental modes to within a few percent accuracy using an expected stellar mass of $M = 1.5M_\odot$. However, it is possible to explain all observed frequencies using only crust shear modes if ρ_p is sufficiently small. As shown in the right panel of Fig. 2, the observed frequencies of 18, 26, 30, and 92.5 Hz can be identified as the $\ell = 3$, 4, 5, and 16 fundamental shear modes, again within a few percent accuracy, assuming a stellar mass of $M = 1.5M_\odot$. These ideas may be important in explaining the observed QPO frequencies in giant flares and the use of asteroseismology in neutron stars may offer new insights into this phenomenon.

3 Conclusion

We consider the effect of nonuniform nuclear structure, the so-called "pasta," in the neutron star crust on torsional shear modes. Based on the suggestion that the elastic properties in the pasta phase could be liquid crystals, the frequencies of shear modes are calculated with a simple function for the shear modulus. As a result, due to the existence of the pasta phase, the frequencies are lower than those expected in the absence of the pasta phase. This result indicates not only the importance of taking the pasta phase into account, but also opens the possibility of probing the pasta structure using observations of stellar oscillations, such as the QPO frequencies in giant flares. Furthermore, we show the possibility of explaining the observed QPO frequencies using only crust shear modes with the pasta phase, which is quite difficult to do in the traditional identification without the pasta phase. In order to find the most suitable stellar model to compare with the observations, we need to adopt a more realistic pasta structure and to examine the systematics with different EOSs.

Acknowledgements This work was supported by Grant-in-Aid for Scientific Research on Innovative Areas (23105711).

References

1. Watts, A.L., Strohmayer, T.E., Adv. Space Res. **40**, 1446 (2006)
2. Sotani, H., Kokkotas, K.D., Stergioulas, N.: MNRAS **375**, 261 (2007)
3. Sotani, H., Kokkotas, K.D., Stergioulas, N.: MNRAS Lett. **385**, L5 (2008)
4. Sotani, H., Colaiuda, A., Kokkotas, K.D.: MNRAS **385**, 2161 (2008)
5. Sotani, H., Kokkotas, K.D.: MNRAS **395**, 1163 (2009)
6. Pethick, C.J., Potekhin, A.Y.: Phys. Lett. B **427**, 7 (1998)
7. Sotani, H.: MNRAS Lett. **417**, L70 (2011)

Whole Earth Telescope Observations of the DAV EC14012-1446

J.L. Provencal and The WET Team

Abstract We report on analysis of 308.3 h of high-speed photometry targeting the pulsating DA white dwarf EC14012-1446. The data were acquired by the Whole Earth Telescope (WET) during XCOV28 in 2008. The Fourier transform of the complete light curve contains 19 independent frequencies distributed in 13 pulsation modes, with dominant peaks at 1,633.9, 1,887.4, and 2,504.9 μHz, as well as numerous combination frequencies. Our analysis reveals that these identified frequencies are consistent with a series of consecutive modes of spherical degree $l = 1$ with an average period spacing of 41 s. Building on this foundation, we calculate preliminary nonlinear fits to high signal-to-noise light curves with the goal of extracting EC14012-1446's convective parameters τ_0 and N.

1 Introduction

Asteroseismology is the study of stellar structure as revealed by global oscillations. The oscillations allow a view beneath the photosphere, and contain information about physical parameters, such as mass, rotation rate, internal transition profiles, compositional structure, and convective energy transport. This information (see for example [2, 4, 6] provides important constraints on fields ranging from stellar formation and evolution, galactic chemical evolution, the age of the galactic disk, the physics of Type Ia supernovae, neutrino physics, and energy flow in stellar atmospheres.

J.L. Provencal (✉)
DARC, University of Delaware, Newark, DE 19716, USA
e-mail: jlp@udel.edu

The WET Team
Delaware Asteroseismic Research Center
e-mail: darc@physics.udel.edu

Provencal, J.L. and The WET Team.: *Whole Earth Telescope Observations of the DAV EC14012-1446*. Astrophys Space Sci Proc. **31**, 277–280 (2013)
DOI 10.1007/978-3-642-29630-7_51,

White dwarfs (WDs) are particularly important targets for asteroseismology. What WDs lack in absolute luminosity they make up for by (1) large amplitudes (0.1–10% flux variations), (2) a rich spectrum of pulsation modes, and (3) a simple underlying structure which makes them easier to model. Pulsating WDs are otherwise normal, so what we learn about their structure can be applied to the entire population of stellar remnants, and further applied to our understanding of their progenitors.

We focus on combining WD asteroseismology with nonlinear analysis of their light curves to provide an empirical description of stellar convection. Montgomery et al. [2] describe convective light curve fitting, a powerful tool using observed nonlinearities in observed WD light curves to probe the physics of WD atmospheres. We present here our results for the DAV pulsator EC14012-1446. Our observational goals were twofold: determine accurate frequency and (l, m) identifications, and obtain several nights of high quality light curves for nonlinear analysis. We present the resulting data set, including the Fourier transform, the dominant frequencies and their (l, m) identifications, preliminary light curve fits, and the current convective parameters for EC14012-1446.

2 Observations and Fourier Transforms

EC14012-1446 (WD1401-147, B=15.67), is a high amplitude, multiperiodic DAV pulsator discovered by Stobie et al. [5]. Our XCOV26 observations span March 25 to April 29 2008. Twenty-seven telescopes distributed around the globe participated in XCOV26, fifteen of which contributed a total of 71 runs. The final XCOV26 light curve contains 308.3 h of high speed photometry. The Fourier transform (FT) of the complete light curve is given in Fig. 1. We identify 19 independent frequencies and numerous combination frequencies (Fig. 1). The spherical indices l, k, and m are derived from an analysis of the combination frequencies and seismology [3].

3 Nonlinear Analysis

Montgomery et al. [2] give a detailed description of fitting observed light curves of WDs to extract the time-averaged thermal response timescale of the convection zone. As first pointed out by Brickhill [1], a pulsating WD's photospheric flux is delayed and attenuated relative to the flux entering at the bottom of the convection zone (cz) by an amount that depends on the cz's thickness. The thermal response timescale, τ_c, is directly related to the convection zone's mass and depth, and so describes its heat capacity. A DAV such as EC14012-1446 will experience local temperature excursions of ± 200 K during a pulsation cycle of 612 s. In response, the cz's heat capacity will vary by a factor of ≈ 10 throughout a pulsation cycle. This process is the source of the large nonlinearities in WD light curves. We focus on

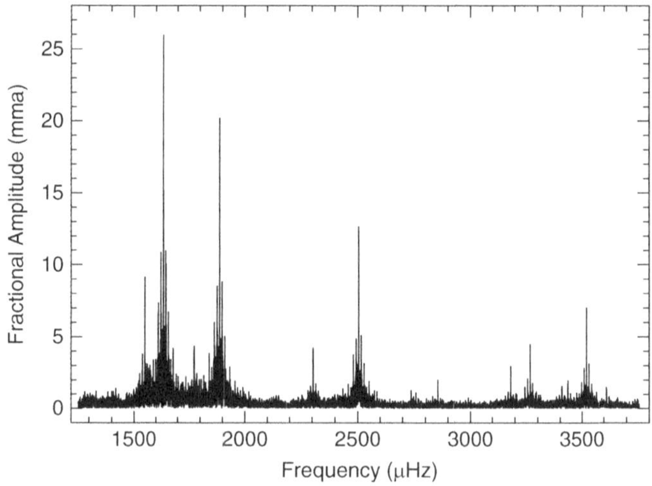

Fig. 1 Main region of power in the EC14012-1446 2008 FT. We find 18 independent frequencies distributed in 13 modes. Peaks above $\approx 3,000\,\mu$Hz are combination frequencies

these observed nonlinearities to determine the time-averaged convective timescale τ_0, the temperature sensitivity parameter N, and the inclination of the pulsation axis to the line of sight θ. Complete mathematical details can be found in Montgomery et al. [2].

We began the fitting process by including the frequencies with firm (l, m) identifications. We then included the additional frequencies with strong (l, m) constraints [3]. This fitting process is *nonlinear.* Adding additional frequencies based on criteria such as amplitude is not necessarily the best procedure. After much experimentation, the final simultaneous fit to high S/N light curves (Fig. 2) produced $\tau_0 = 99.4 \pm 12$ s, $\mathrm{N} = 85 \pm 6$, $\theta = 32°.9 \pm 3.2$.

4 Summary

Until the implementation of convective light curve fitting, our Sun was the only star for which empirical information on its convection zone could be obtained. Determinations of the time averaged convective timescale τ_0 and the temperature sensitivity parameter N can now potentially be obtained for any white dwarf pulsator of moderate amplitude. We have taken the first steps with EC14012-1446.

Approximately two-thirds of all white dwarf pulsators show significant nonlinearities in their light curves. Our ultimate goal is to map τ_0 as a function of $T_{\mathrm{eff},0}$ and $\log g$ for a population spanning the instability strips of both the hydrogen atmosphere DAVs and the helium atmosphere DBVs. Such a map will provide important empirical constraints on convection for white dwarfs and eventually other types of pulsating stars.

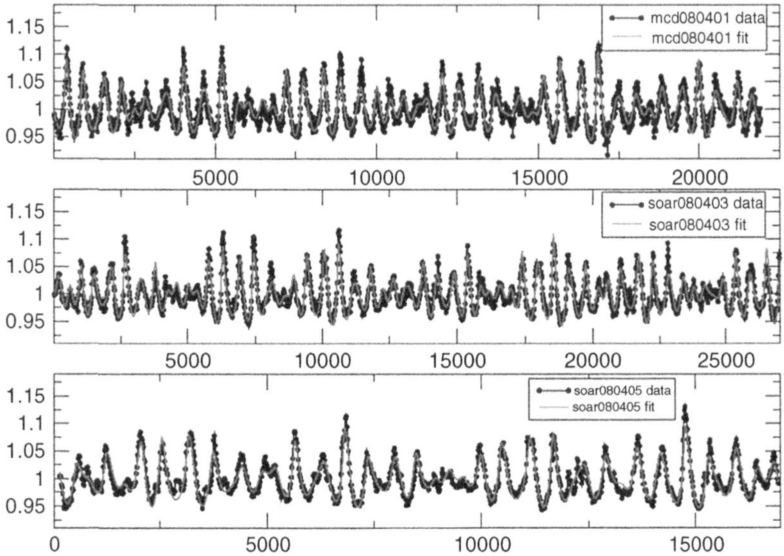

Fig. 2 Simultaneous fit of EC14012-1446's frequencies (*solid line*) to the light curves (*filled circles*). Note the change in x axis for each plot

Acknowledgements Space limitations force us to limit the number of individual authors who contributed to this work. These WET observations could not happen without the enthusiastic support of everyone in the network.

References

1. Brickhill, A.J.: MNRAS **259**, 519 (1992)
2. Montgomery, M.H., Provencal, J.L., et al.: ApJ **716**, 84 (2010)
3. Provencal, J.L.: ApJ **751**, 91 (2012)
4. Stello, D., et al.: ApJ **738**, 28 (2011)
5. Stobie, R.S., et al.: MNRAS **272**, L21 (1995)
6. Winget, D.E., et al.: ApJ **430**, 839 (1994)

Part VII
Conference Closing Remarks

Closing Summary: Twentieth Stellar Pulsation Conference

Joyce Ann Guzik

Abstract There have been many significant advances in the 40 years since the first stellar pulsation conference in 1971, including new instrumentation and observational techniques, discovery of many thousands of variable stars from extensive surveys, and prediction, discovery, and explanation of many new types of variable stars. These advances have generated new insights into the physics required to understand and interpret stellar phenomena, and motivated progress in computational capabilities. However, with increased resolution and sensitivity, data volume and accuracy, and the ability to include more complex processes in computational models, we also have new challenges and questions. This conference pointed to some emerging directions for the study and application of stellar pulsations. Here we summarize a few of the new discoveries, observational capabilities, progress in theory and computational models, and questions discussed.

1 New Variable Star Types Predicted

While the ϵ mechanism for pulsation driving has been discussed at previous pulsation conferences, this mechanism has yet to be confirmed as a cause of pulsations in any type of variable star. Here we heard that the ϵ mechanism may drive pulsations in the ^{3}He-burning region of very low metallicity (Z=0.0001) stars (but can we find such stars today?), and in the deuterium and ^{3}He-burning-regions of low-mass M dwarf stars. We also learned that the convective blocking mechanism that drives pulsations in γ Dor stars may also drive pulsations for partially convective 0.4–0.5 $M_\odot$ M dwarfs, and for stars of higher mass but low metallicity. If found, these new types of pulsating stars would further fill out our

J.A. Guzik (✉)
Los Alamos National Laboratory, XTD-2, MS T-086, Los Alamos, NM 87545, USA
e-mail: joy@lanl.gov

Guzik, J.A.: *Closing Summary: Twentieth Stellar Pulsation Conference*. 283
Astrophys Space Sci Proc. **31**, 283–288 (2013)
DOI 10.1007/978-3-642-29630-7_52,

HR diagram that contains pulsating variables at nearly every stellar mass and evolutionary state.

2 2D/3D Stellar Models and Databases

While up to recent times most stellar evolution and pulsation studies have used 1D stellar models, we learned about new 2D and 3D model codes being applied to pulsation problems (SPHERLS, ANTARES, ACOR, ROTORC, and LNRO).

Studies of 2D polytropic rotating models first highlighted the inadequacies of categorizing the pulsation modes using spherical harmonics. We learned that there has been a great deal of progress in characterizing and looking for patterns in the predicted stellar pulsation spectrum modified by rotation, so dealing with this complication may not be as dismaying as it seemed a few years ago.

We also learned about the need for massive databases with the new surveys and finding more efficient automated ways to do asteroseismology. The VOTA (Virtual Observatory Tool for Asteroseismology) database project has 500,000 intermediate-mass stellar models so far, and there are plans to link to other databases.

3 Solar-Like Oscillations in Red Giants

About a decade ago, we were still debating whether red giant Mira variables were pulsating in the fundamental or first overtone mode, and there seemed to be little prospect of using red giant pulsations to infer much about the evolution history or internal structure of these stars. One of the most surprising and valuable advances has been the discovery and exploitation of solar-like oscillations in red giants using *Kepler* and CoRoT data. The frequency separations of the p modes and mixed p- and g-type modes have been used to infer that the interior rotation rates are 3–10x the surface rate. $\Delta\nu$ vs. frequency diagrams are even being used to separate red giant populations of different masses and evolutionary origin (e.g., core He burning, H burning, etc.) and to measure He core masses.

4 Classical and Exotic Pulsators

We continue to discuss and make progress in the study of the classical large-amplitude Cepheid and RR Lyrae variables that are pulsating most obviously in one or two radial modes, although the probability of additional nonradial modes has been suggested for these stars. Period doubling has been discovered in many *Kepler* Blazhko stars. A resonance between the fundamental and the ninth overtone that is a strange mode has been suggested as the cause of this period doubling.

Cepheids with low-mass companions have a different period distribution than those with higher-mass companions, and the study of Cepheids with low-mass binary companions will be important for study of star formation. The mass-loss rates of δ Cephei may have been measured at $\sim 10^{-6}$ $M_\odot$/year, and mass loss in Cepheids may be connected to pulsations. Cepheid period-luminosity relations, critical for setting the cosmic distance scale, are tighter/more accurate at certain pulsation phases in the V band.

The first detection of solar-like oscillations in a δ Scuti star was announced by Antoci et al. (to appear in it Nature).[1] Amplitude and phase variability introduces new frequencies into the spectrum of high-amplitude δ Scuti stars. New pre-main sequence (PMS) δ Scuti stars, and the first PMS γ Dor star were discovered in the cluster NGC 2264. No pulsations have been confirmed in Vega (yet).

Pulsating proto-planetary nebula stars are expected to change periods at a rate of 3 days over 15 years–such a high period change rate should be observable, but hasn't been seen so far. Whole Earth Telescope observations of DA white dwarf pulsations are being used to probe the properties of the convection zone in white dwarfs. Quasi-periodic oscillations observed in neutron stars may be explained by torsional shear modes, and may be useful to diagnose properties of the neutron star crust.

We also heard various perspectives on the Be-star phenomenon. Prograde modes transfer angular momentum and may cause mass loss in Be stars. But it is possible that pulsation is coincidental and plays no role in the Be-star emission. There was debate on whether g modes are actually present in Be stars. Rotation and magnetic fields were also discussed as a cause of mass ejection, but there is no evidence for strong dipole B fields.

5 *Kepler* sdB and White Dwarf Stars

Thirteen g-mode pulsating subdwarf B stars have been discovered in the *Kepler* data. Seventeen percent of the 317 modes seen are $l = 1$ and 2. Mode trapping is softened by diffusion in these stars. We may be able to understand the channels of origins of sdB stars using period spacings, analogous to the studies of red giants using solar like-oscillations. The sdB stars have the same core as in red giants, but don't have the extended atmospheres.

A DB white dwarf pulsator was discovered in the *Kepler* field, so it will be possible take advantage of the high signal-to-noise *Kepler* photometry for a white dwarf after all, and maybe to measure its evolutionary period change if the star can be observed continuously for a few years.

---

[1]Editors' note: Due to Nature publication policy, the presentation by Antoci et al. has been not included in these proceedings book. We refer the reader to the Nature's publication reference: Antoci et al.: Nat 477, 570 (2011)

6 Stellar Pulsation to Test Stellar Evolution

We learned from A. Weiss's overview that stellar evolution modelers would like help from stellar pulsation experts to test and constrain their models. In particular, they would like to know the extent of mixed cores, composition profiles, internal rotation profiles, diffusion constraints, abundance mixtures (to resolve the solar abundance problem), helium abundances, cluster ages, and period changes.

7 New Observational Capabilities

We learned that there is promise/hope of future data and life after *Kepler* and CoRoT in the planned space missions PLATO and BRITE, in ground-based networks such as SONG, long-baseline interferometry (CHARA and VLTI) and even spectro-interferometry (VEGA/CHARA), IR observations, and high-resolution spectroscopic surveys. The VVV (Vista Variables in the Via Lactea) project will measure one million light curves of stars toward Galactic bulge over next 4 years in the infrared. Interferometry can measure Cepheid radii and distances directly, as well as examine circumstellar disks, winds, and roAp star properties. Interferometry will be used to measure the stellar diameters of 400 stars brighter than 6.5 mag to within 2%. Interferometry combined with radial velocity measurements will obtain distances to 22 Cepheids.

As discussed in the closing session, we must be vigilant to advocate for resources for ground-based followup observations of stars observed in space missions to derive fundamental properties, confirm planet detections, and for longer-term monitoring. The space-based or single-instrument data will not be useful without a combination of diagnostics to test hypotheses and provide constraints.

8 Surface Effects

The treatment of stellar surfaces creates difficulties in interpreting stellar data, and more refinement of stellar model physics will be necessary to compare to observed stellar frequencies and amplitudes. It was interesting to learn the extent to which photometric mode identification depends on the physics of modeling stellar surfaces, including convection treatment, model atmospheres, and opacities. We learned by J.C. Suárez that rotation may be just as important as turbulence or diffusion in contribution to the surface-effect corrections that are applied to the adiabatic pulsation frequencies in sun-like stars. A time-dependent convection treatment was evaluated and proposed to account for these surface-effect corrections.

9 Synergy Between Pulsation, Activity, and Planet Studies

Asteroseismology is improving constraints on stellar masses, radii, compositions, and ages, and hence informing planet formation, constraining stellar evolution models, and realizing its promises in these areas. Asteroseismology (if the error bars are believable) is refining planetary masses and ruling out some objects as planets. Lithium depletion and higher metallicities are observed for stars with planets, providing an interesting clue to planet formation. The rotation axis of the star HR8799 is misaligned with the axis of its planetary orbits–how did this system form?

10 Stellar Activity Cycles

We learned a great deal about stellar activity cycles at this conference, and how stellar pulsations can help diagnose stellar activity. This is an important development as we don't fully understand and can't predict aspects of the activity cycle of our Sun. Clues from other stars will be important to developing this understanding, and to learning whether the Sun is unusual or typical.

Stellar time series data can be used to follow spot formation and migration, and is being used successfully to test dynamo theory. Short stellar activity cycles are more common than we once thought. We learned that the solar twin 18 Sco has a 7-year activity cycle, and perhaps Jupiter-sized planets and a 23-day rotation period. Scaling of solar oscillation amplitudes to other stars is not working as well as expected, and activity may have something to do with the discrepancies. Acoustic mode activity signatures are present even when there is no evidence of spots. There appear to be two branches of the ratio of the rotation period to the stellar activity cycle period, but the origin of this division is not known.

11 Issues Discussed

Several issues were discussed during the meeting:

1. What is the nature of the low-frequency modes (<0.5 cycle/day) seen in A-F stars in the *Kepler* and CoRoT data and elsewhere? Do these have anything to do with pulsations? Or do they result from mode combinations, rotation, activity, mode coupling?
2. Is asteroseismology delivering on its promises? Are our advancements worthy of the investments we have made so far? (The answers generally were that we have made a great deal of progress, but we also have much more to do.)

3. Are the error bars published on mass, radius, metallicity, age, etc. derived asteroseismically by our model grid searches too small? Should they be caveated more explicitly?
4. What are we going to do about our collective "headache" from the new categories of pulsation modes created by rotation?
5. What is the role of pulsation in the Be-star phenomenon?

12 Interdisciplinary Theme

The organizers endeavored to make this meeting interdisciplinary and introduce new research directions by including talks covering stellar evolution, stellar activity, and planets. The stellar pulsation meetings have always been to some degree interdisciplinary, including applications of stellar pulsation, such as calibrating the Cepheid distance scale for cosmology, and discussing how stellar pulsations inform the physics of stellar interiors that are also applicable in other fields. Such physics includes opacities, equation of state, nuclear reactions, turbulence modeling, and chemical diffusion. At this meeting we have heard from stellar evolution experts about what is needed interpretations of stellar pulsations (e.g. abundance gradients, internal rotation profiles, ...). We learned that there is synergy between study of pulsations and the study of stellar activity, stellar cycles and planet transits, and it is necessary to consider all of these processes together to interpret observations. Also, the Be star controversy, and whether the Be star phenomenon is caused or affected by pulsations, or has something to do with rapid rotation or magnetic fields, was framed and debated.

In many previous pulsation meetings a session was devoted to each type of pulsating variable star, making it difficult for researchers on one type of variable to find some connection to other types. At this conference perhaps for the first time the division between these types was not as apparent, and the connections happened naturally instead of being contrived. We think this is a positive development, as working together will be essential to make progress in dealing with many formidable challenges that are emerging in interpreting pulsation data.

13 Next Conferences

Andrzej Pigulski announced that the next (21st) stellar pulsation conference will be held August 18–23, 2013 in Wroclaw, Poland. Wolfgang Gieren proposed to host the 22nd conference sometime between November 2015 and March 2016 in Chile.

Acknowledgements The conference participants acknowledge the entire Local and Scientific Organizing committees for their support, dedication and service to our community in creating an outstanding meeting in the beautiful location of Granada.

Part VIII
Poster Contributions

The poster contributions are presented as electronic supplementary material.

List of Poster Contributions

Electronic supplementary material The online version of this article (doi:10.1007/978-3-642-29630-7_53) contains supplementary material, which is available to authorized users.

First author	Title	Identifier
A. Arellano Ferro	New Blazhko variables in the globular cluster NGC 5024	1
A. Frasca	Magnetic activity and differential rotation in very young stars in the Kepler FoV. The case of KIC 8429280	2
A. Kaiser	Stellar characterization of large spectroscopic datasets: A bayesian approach	3
C.C. Ngeow	Ground-based BVRI follow-up observations of two Cepheid candidates in Kepler's field	4
C. Rodríguez-López	Pulsations in M dwarf stars	5
D. Bramich	Exploring the variable stars in NGC 6981 and NGC 5024: RR Lyr and SX Phe	6
D. Citro	Random phase corrections to the IRAC bands PL relation using SAGE data for the LMC/SMC	7
D. G. Turner	Recent observations of HDE 344787, the smallest amplitude classical Cepheid	8
D. Mozdzierski	Variability survey in the young open cluster NGC 457	9
E. Brunsden	Spectroscopic pulsations of γ Doradus stars	10
E. Guggenberger	The changing Blazhko effect of KIC 6186029: a comparison with other cases of non-repetitive Blazhko cycles	11
E. Kanev	Radii of LMC & SMC Cepheids	12

(continued)

Suárez, J.C. et al. (eds.).: *List of poster Contribution.*

Astrophys Space Sci Proc. **31**, 291–293 (2013)
DOI 10.1007/978-3-642-29630-7_53, © Springer-Verlag Berlin Heidelberg 2013

First author	Title	Identifier
E. Bellinger	New insights into the mean light Cepheid PL relation through the use of multiphase relations	13
E. Plachy	Detection of chaos in RR Lyrae models	14
F. Ekmekçi	RR Lyr type stars, ST Boo and RR Leo: 2007 observations & preliminary results of the frequency analysis	15
G. Kopacki	P-L relation for SX Phoenicis stars from galatic globular clusters	16
	Variable stars in the globular cluster M79	17
H. Sotani	Signatures of hadron-quark mixed phase in gravitational waves	18
H. Neilson	Constructing a Cepheid period p-factor relation from static stellar atmospheres	19
J.H. Peña	Membership determination of the variable stars in the direction of the open cluster NGC 6811 through Strömgren photelectric	20
J. Molenda-Zakowicz	Spectroscopic study of two open clusters in the Kepler FoV: NGC 6811 & NGC 6866	21
	Standard photometry of Kepler asteroseismic targets	22
J. Fernández	Mode properties of δ Scuti V830 Her & HD 163032 with the Taiwan Automated Telescope Network	23
K. Uytterhoeven	The SONG project and the prototype node at Tenerife	24
L. Çelik	How to *correctly* stitch together data of a Blazhko star?	25
M.E. Escobar	Detailed asteroseismic modeling of the CoRoT main target HD 522265	26
M. Skarka	Physical parameters of RR Lyrae type stars DH Peg & UY Cam	27
R. Medupe	A formula for mode-identification	28
G. Michalska	Variable stars in the open cluster NGC 6611	29
M. Jerzykiewicz	Baade-Wesselink Radius and distance of V1154 Cygni	30
S. Ocando	δ Scuti and γ Doradus variables in open clusters	31
P. De Cat	Spectral disentangling and pulsational analysis of the double-lined binary HD 147787	32
P. Lenz	KIC 9700322 revisited: Testing the binary star hypothesis	33
P. Moskalik	Discovery of peculiar double-mode pulsations and period doubling in Kepler c-type RR Lyrae stars	34
P. A. Bradley	Analysis of γ Doradus & δ Scuti stas from the Kepler satellite	35
P. R. Wood	Variability properties of Magellanic clould post-AGB candidates	36

(continued)

First author	Title	Identifier
P. Kervella	High accuracy interferometric radii of asteroseismic targets: a powerful constraint for stellar models	37
R. Samadi	Amplitude of acoustic modes in red giant stars: departure from quasi-adiabatic treatment	38
S. Martín-Ruiz	Is V994 Her really a quadruple system of two eclipsing binaries?	39
S. Prins	MAIA, a fast three-channel imager for asteroseismology	40
S. Théado	Iron accumulation, thermohaline convection and stellar oscillations	41
S. Morgan	Cepheid instability strip "first-time crossers"	42
R. Smolec	Non-local model for the turbulent fluxed due to thermal convection in rectilinear shearing flow	43
S. Reyner	The approximation of RR Lyrae and eclipsing binary light curves using cubic polynomials	44
W. Szewczuk	Constraints on model atmospheres from complex asteroseismology of the β Cephei stars	45
	Studying the hybrid pulsators 12 Lac: extended mode identification and complex seismic modelling	46
C. Ulusoy	New studies of a variety of Southern pulsating B stars	47
V. Maestro	Interferometric observations of rapid rotators	48
Z. Bognár	Investigation of rotational splitting in the pulsating white dwarf GD 154	49

The manufacturer's authorised representative in the EU is Springer Nature Customer Service Centre GmbH, Europaplatz 3, 69115 Heidelberg, Germany. If you have any concerns regarding our products, please contact ProductSafety@springernature.com

Printed and bound by CPI Group (UK) Ltd, Croydon, CR0 4YY
30/09/2026
02225595-0016